USB
Peripheral Design

John W. Koon

Annabooks

San Diego

"John Koon has taken on a formidable task...providing design information to a developer community in an industry that is changing rapidly. He is to be commended. John has put together a great starting point for anyone who is beginning a USB peripheral design project."

Mark Richman, USB Product Line Manager,
Microelectronics Group, Lucent Technologies

"We look forward to the proliferation of USB peripherals in the marketplace in 1998. We have many years of experience in developing the technology, and also serve as the Chair of the USB Audio Device Class. Although we are in the audio business, we believe that the USB market will develop through the availability of a variety of USB peripherals. An initiative like John's will serve to bring everything together and advance the USB initiative."

Achiel Verheyen, Business Development Manager
Philips MM Speaker Systems

"Koon provides an excellent introduction and elaboration of the issues that peripheral developers will need to solve in order to achieve compatibility with the USB universe."

Barry Hoberman, Senior Director, Virtual Chips Products
Phoenix Technologies Ltd.

USB
Peripheral Design
by
John W. Koon

PUBLISHED BY

Annabooks
11838 Bernardo Plaza Court
San Diego, CA 92128-2414
USA

619-673-0870
800-462-1042
619-673-1432 FAX
info@annabooks.com
http://www.annabooks.com

Copyright © John W. Koon 1998

All rights reserved. No part of the contents of this book may be reproduced or transmitted in any form or by any means without the prior written consent of the publisher, except for the inclusion of brief quotations in a review.

Printed in the United States of America

ISBN 0-929392-46-9

First Printing February 1998

Information provided in this publication is derived from various sources, standards, and analyses. Any errors or omissions shall not imply any liability for direct or indirect consequences arising from the use of this information. The publisher, authors, and reviewers make no warranty for the correctness or for the use of this information, and assume no liability for direct or indirect damages of any kind arising from technical interpretation or technical explanations in this book, for typographical or printing errors, or for any subsequent changes.

The publisher and authors reserve the right to make changes in this publication without notice and without incurring any liability.

All trademarks mentioned in this book are the property of their respective owners. Annabooks has attempted to properly capitalize and punctuate trademarks, but cannot guarantee that it has done so properly in every case.

About the Author

John W. Koon has been active in USB since its beginning. He is a frequent speaker at technical events and developer's conferences, and is the author of a number of papers and articles that have appeared in a variety of publications. John has a BSEE from California State Polytechnic University, Pomona, and an MBA from San Diego State University. He lives in San Diego, California with his wife and two children.

Dedication

For my beloved wife Carol and our children Hannah and Nathan

Acknowledgments

I want to thank the following individuals for their contribution and review various parts of the book: Achiel Verheyen of Philips, Abdul Jarrar of Intel, Barry Hoberman of Phoenix Technologies, Dave Bursky of Electronic Design Magazine, David Etzkorn of Huntsville Microsystems, Dennis Chang of Alcor Micro, Estella Siau, Jeffrey Goodwin of Ashley Laurent, Jonathan Bearfield of Texas Instruments, Kevin Lynn of Micrel, Mark Richman of Lucent Technologies, and the staff at Annabooks.

Special thanks go to the following groups who have been very instrumental in the completion of this book:

USB-IF: Steven Whalley, Chairman of the USB Implementer's Forum and Connectivity Initiatives Manager of Intel, and Ralph S. Smith, Technical Marketing Engineer for their support and permission to quote from the USB specification.

Microsoft: Carl Stork, General Manager of PC Hardware Strategy, Mark Williams, Windows PC Hardware Evangelist, and Clark Sealls, Program Manager of USB Development for their help with the review of the Windows® 98 software section to assure its accuracy.

Finally, my sincere appreciation goes to Agnes M. Chin who kindly provided the cartoons used in this book.

Contents

1.	**INTRODUCTION**	**1**
1.1	Purpose of this Book	1
1.1.1	Key points in this chapter	1
1.1.2	Prerequisites	1
1.2	Definition of Universal Serial Bus (USB)	2
1.3	Goals of USB	6
1.4	History of USB	6
1.5	USB Products Supported	7
1.6	Market Potential and Opportunities for USB Peripheral Developers	8
2.	**ORGANIZATION OF THIS BOOK**	**11**
2.1	Purpose of this Chapter	11
3.	**USB SYSTEM COMPONENTS OVERVIEW**	**13**
3.1	Purpose of this Chapter	13
3.1.1	Key points in this chapter	13
3.2	Summary of USB Features	14
3.3	Back Panel of the USB PC	15
3.4	Description of USB System Components	17
3.4.1	Software Components and Windows 98	17
3.4.2	Hardware Components	25
3.5	Review of the Interlayer Communication Model	35
3.5.1	On the Host side	36
3.5.2	On the Device side	36

3.6	USB System Operation Summary	37
3.7	Application of the Model	37

4. EXAMPLES OF USB DEVICES 39

4.1	Purpose of this Chapter	39
4.2	Cables and Connectors	40
4.3	Communications Interface	43
4.4	Computer Attached Telephone	44
4.5	Digital Video Camera	45
4.6	Gamepad	45
4.7	Hand Input Device	46
4.8	Host Systems	48
4.9	IR Remote Control	52
4.10	ISDN Controller	53
4.11	Keyboard	55
4.12	Motherboards	58
4.13	PC-to-PC Link	59
4.14	PCI-to-USB Controller Board	60
4.15	Printer Cable	61
4.16	Remote Control Device	62
4.17	Scanner	63
4.18	Speakers	64

4.19	Stand-alone Hub	66
4.20	USB-to-ECU Bridge	67
4.21	USB-to-RS232 Bridge	68

5. DESIGNING USB HUBS AND DEVICES 71

5.1	Purpose of this Chapter	71
5.1.1	Key points in this chapter	71
5.2	Design Goals and Trade-offs	72
5.3	Driver Issues	72
5.3.1	Generic Device Drivers Available in Microsoft's Windows 98	72
5.3.2	Write Your Own DD	73
5.3.3	Buy the DD and Firmware	73
5.3.4	Custom DD	73
5.4	USB Hub Design Considerations	74
5.4.1	Basic Hub Functionality	74
5.4.2	Hub Feature Considerations	75
5.4.3	Hub IC Component Selection Criteria	75
5.5	USB Device Design Considerations	77
5.6	Device IC Components Selection Criteria	79
5.7	Power Considerations	80
5.7.1	Power Management Design considerations	81
5.8	Compatibility and Testing Issues	85
5.8.1	Cables and Connectors	85
5.8.2	Other Issues	86
5.8.3	Mechanical, Industrial Design, and Prototyping	87
5.9	Custom Application Specific Integrated Circuits (ASIC)	87

6. EXAMPLES OF USB HUB AND DEVICE ICS 89

6.1	Purpose of this Chapter	89

6.2	Device ICs	91
6.3	Host ICs	121
6.4	Hub ICs	130
6.5	Power ICs	159
6.6	Other ICs	163

7. USB DEVELOPMENT TOOLS AND HELPFUL SOURCES — 173

7.1	Purpose of this Chapter	173
7.2	Bus Analyzer	175
7.3	Custom Device Drivers	178
7.4	Hardware Emulator	180
7.5	Input Device Developer's Kit	180
7.6	Host Controller Developer's Kit	181
7.7	Monitor Firmware	183
7.8	OS Utility	187
7.9	Overcurrent Protection	190
7.10	Printer Firmware	191
7.11	Product Development Services	195
7.12	Protocol Analyzer	195
7.13	Silicon Design Services	197
7.14	Simulation Models	198
7.15	Synthesizable Cores	204

7.16	Transceiver Design		231
7.17	Workshops, Conferences, and Books		232

8. APPENDIX — 233

8.1 USB Power Application Notes — 233
- 8.1.1 Purpose of this section — 233
- 8.1.2 Introduction — 234
- 8.1.3 Classes Of Devices — 235
- 8.1.4 Protection Requirements — 236
- 8.1.5 Connectivity Limitations — 238
- 8.1.6 Power Supply Voltages — 239
- 8.1.7 Voltage Requirements — 240
- 8.1.8 Current Limiting Devices and Power Switches — 243
- 8.1.9 Conclusion — 246
- 8.1.10 References — 246

8.2 USB Power Management — 247
- 8.2.1 Abstract — 247
- 8.2.2 Introduction — 247
- 8.2.3 Power management — 248
- 8.2.4 Device Classes — 249
- 8.2.5 USB Power Distribution — 251
- 8.2.6 Self-Powered Hub Requirements — 253
- 8.2.7 Integrated High-Side Power Switches — 257
- 8.2.8 Transient Droop — 258
- 8.2.9 Layout — 258
- 8.2.10 Bus-Powered Hubs — 268
- 8.2.11 Bus-Powered Hub Requirements — 269
- 8.2.12 Discrete Power Switch — 272
- 8.2.13 Soft start — 272
- 8.2.14 USB Cables — 274
- 8.2.15 Detachable Cables — 275
- 8.2.16 Voltage Drop — 276
- 8.2.17 Layout — 277
- 8.2.18 Micrel Power Switches — 279
- 8.2.19 Summary — 282

8.3 Glossary — 283

8.4 Frequently Asked Questions (FAQ) — 296

8.4.1	What is USB?	298
8.4.2	What kind of peripherals will USB allow me to hook up to my PC?	298
8.4.3	How does it work?	298
8.4.4	Will I need to purchase special software to run USB peripherals?	299
8.4.5	Will USB peripherals cost more?	299
8.4.6	Is there a Mac version of the standard?	299
8.4.7	Are there USB products out right now?	299
8.4.8	How can USB be used between two hosts, like a laptop and a desktop?	300
8.4.9	How does USB compare with the FireWire/IEEE 1394 standard?	300
8.4.10	When it is available, will FireWire replace USB?	300
8.4.11	Who created USB, anyway?	301
8.4.12	What are the intellectual property issues with USB? Is there a license, what does it cost, and what is the "Reciprocal Covenent Agreement" I've heard about?	301
8.4.13	What is the USB-IF?	301
8.4.14	What are the benefits of USB-IF?	302
8.4.15	How do I join USB-IF?	302
8.4.16	How do I get in touch with USB-IF?	302
8.4.17	What about OHCI and UHCI?	303
8.4.18	Where do I get the SIE VHDL?	303
8.4.19	Where can I get a spec?	303
8.4.20	Is there a newsgroup for USB?	303
8.4.21	How do I get a USB vendor ID?	304
8.4.22	Where can we get EMC testing peripherals?	304
8.4.23	Hasn't someone informed the trade press that there is a need for Firewire and USB?	304
8.4.24	Is the USB bus going to have a long distance 50-200 meter extension (possibly fiber) for these large customers that need the capability?	304
8.4.25	Will legacy device support be in the formal USB spec? When?	305
8.4.26	Will source code for driving HCI chips be made available?	305
8.4.27	When a device is detached, its device driver is unloaded. If that device is re-inserted, would its driver be reloaded?	305
8.4.28	Are there any plans to increase the bus bandwidth of USB in the future to 2x, 3x?	305
8.4.29	Can someone clarify the difference and applications for series A and series B connectors?	305

8.4.30	What is the difference between a root hub and normal hub in terms of hardware and software?	306
8.4.31	Is USB a viable bus for peripherals like CD-R, tape, or hard disk drives?	306
8.4.32	The programming spec for UHCI is not publicly available. When can one get the UHCI spec?	306
8.4.33	How do I get a USB PDK system?	307

9. VENDOR LIST — 311

9.1	Unsorted	311
9.2	Integrated Circuits	315
9.3	Power ICs	318
9.4	Tools	318
9.5	Help	320
9.6	Cables	320
9.7	Connectors	321
9.8	Keyboards	321
9.9	Joysticks	322
9.10	Systems	322
9.11	Mice	323
9.12	Printers	324
9.13	Computer Telephones	324
9.14	Video Cameras	324
9.15	Infra-Red	324
9.16	Motherboards	325

9.17	Software Core	325
9.18	Gamepad	326
9.19	USB Host and Function Macros in Netlist or RTL USB Test Environment	326
9.20	Standards	326
10.	**INDEX**	**327**

List of Figures

Figure 1.1a: Simple Host and Device Relationship 2
Figure 1.1b: Master-Slave Relationship 2
Figure 1.2: Relationship of USB and PCI 3
Figure 1.3a: USB Topology with Standalone Hub 4
Figure 1.3b: USB Topology with Monitor Hub 5
Figure 3.1: Summary of USB Features 14
Figure 3.2: The Present USB Back Panel 16
Figure 3.3: The Future USB PC Back Panel 16
Figure 3.4: Host Controller Protocols 26
Figure 3.5: Interlayer Communication Model 27
Figure 3.6a: Series A Cable Plug 28
Figure 3.6b: Series B Cable Plug 29
Figure 3.6c: USB A to B Cable 29
Figure 3.6d: Fully Rated Cable (shielding not shown) 30
Figure 3.6e: Sub-Channel Cable (unshielded) 30
Figure 3.7a: Series A Receptacle 31
Figure 3.7b: Series A Receptacle (stacked) 31
Figure 3.7c: Series A Receptacle (vertical) 32
Figure 3.7d: Series B Receptacle 32
Figure 3.7e: Series B Receptacle (vertical) 33
Figure 3.8: Pig-Tail Cable 33
Figure 3.9: Summary of USB Cable Characteristics 35
Figure 3.10: VESA Enhanced Video Adapter (EVA) Defines
    USB Pin-Outs 35
Figure 5.1: Hub Function 74
Figure 5.2a: Low-Speed Device Cable and Resistor Connection 78
Figure 5.2b: High-Speed Device Cable and Resistor Connection 78
Figure 5.3: Compound Self-Powered Hub 81
Figure 5.4: Compound Bus-Powered Hub 82
Figure 5.5: Self-Powered Function 83
Figure 5.6: High-Power, Bus-Powered Function 84
Figure 5.7: Low-Power, Bus-Powered Function 84
Figure 1. Self-Powered Hub Voltage Drops 260
Figure 2. Stand-Alone or Unregulated Input Self-Powered Hub 264
Figure 3. Self-Powered Hub with Individual Power Switches 265
Figure 4. Ganged Switch Four Port Self-Powered Hub 266
Figure 5. Ganged Output 4 Port Bus-Powered Hub 279

Foreword

Universal Serial Bus (USB) is rapidly becoming the fastest and easiest way to handle peripheral device connectivity in the world of PC's. By the end of this decade, USB will almost certainly be a household name, resident on all PC's and most high volume, commodity peripherals. Looking back at the turn of the next century, one may wonder how we ever did without it for so long.

From its conception in 1994, USB has been embraced by the world's major PC OEMs (Original Equipment Manufacturers), PC peripheral suppliers, silicon, tools, cables and connectors, and software companies. Now, just three years later, many products are appearing in retail stores with yet more creative USB product innovations being announced almost weekly. John's book has captured the core of all these USB building blocks in one easy guide and put them at our fingertips for quick reference to valuable data, whether you are a novice or seasoned veteran of USB.

This book is really the complete reference manual all design, marketing, sales, and product managers should have on their desk. As well as a thorough overview of USB technology and frequently asked questions, it provides all the names and contacts you will need to get you started on USB product development. It also provides a comprehensive list of available USB products that are available for development, testing, retail, or PC 'bundling' use.

These kinds of books do not often appear so early on in the development of a new technology, yet this is just the time when you need them. To compile such a complete and practical guide of new and often unannounced products involves many hours of painstaking searches, legwork, and phone calls around the globe and around the clock! John's tenacity at doing this has resulted in a 'must have' book for anyone involved in USB.

Steve Whalley
Chairman, USB Implementors Forum
Connectivity Initiatives Manager, Intel
Chandler, Arizona

Preface

The Universal Serial Bus (USB) has built-in intelligence in the operating system and peripheral silicon, which makes the host-to-peripheral device connection very easy. Many of the configuration chores that need to be done today are taken care of automatically. Many more generic drivers will be shipped with Windows 98. Other USB features, including hot plug and detachment, are part of the USB specification. The ease of use of the USB, along with the strong support from the industry (the seven proponents include Compaq, Digital Equipment, IBM, Intel, Microsoft, NEC, and Northern Telecom) will accelerate the growth of the computer market, including the home PC. Most of the PC's shipped in the US today already have the USB port built-in. In the next eighteen months we will witness the introduction of many USB peripheral products. The 400-plus member USB-IF (USB Implementers Forum) has put forth major effort to educate and promote the USB technology at both the OEM and retail level.

The key to success for a developer or OEM product integrator in this market is the ability to bring the USB products to market in a timely fashion. It is paramount to have access to information on the USB building blocks such as software, silicon, development tools, training, and consulting help. This book is written to provide developers and OEM product integrators with this information in one publication. Additionally, the products mentioned will have a contact address and web site if further information is needed. For new product updates or free product registration contact http://www.usbnews.com.

John W. Koon
San Diego, California

Before USB

1. Introduction

1.1 Purpose of this Book

This book is a complementary publication to the Universal Serial Bus (USB) Specification Version 1.0, January 19, 1996. While the focus of the USB Specification is on the theoretical modeling and requirements, this book is on the implementation of the USB Specification.

There are two major objectives of writing the Design Guide.

- To provide useful information to facilitate the design of USB Peripheral Products.
- To provide examples of USB products available to the USB OEM system integrators.

1.1.1 Key points in this chapter

Prerequisites

Definition of Universal Serial Bus (USB)

Goals of the USB

History of the USB

USB Products supported

Market potential and opportunities for USB peripheral developers

1.1.2 Prerequisites

You are encouraged to first read and be familiar with the Universal Serial Bus (USB) Specification Version 1.0, January 19, 1996. A copy of the specification can be downloaded from http://www.usb.org.

1.2 Definition of Universal Serial Bus (USB)

USB is a peripheral bus standard originally developed by seven core companies (Compaq, DEC, IBM, Intel, Microsoft, NEC, and Nortel) to provide a simple way to integrate computer and telephone, commonly known as CTI (computer-telephone-integration). It later expanded into a concept of easy integration of multiple (up to 127) peripherals to the host Personal Computers (PCs). USB operates under a master/slave scheme and each peripheral talks to the host either directly (see Figures 1.1a and b) or via a hub (see Figures 1.3a and b). It is important to understand that USB is a serial bus topology, not to be confused with the conventional RS-232 serial interface. In addition, it doesn't replace the PCI bus, but works with it (see Figure 1.2).

Figure 1.1a: Simple Host and Device Relationship

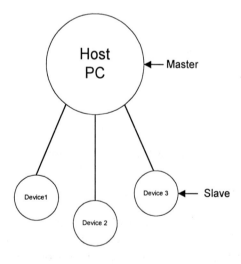

Figure 1.1b: Master-Slave Relationship

Chapter 1: Introduction

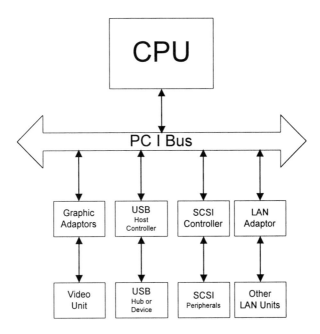

Figure 1.2: Relationship of USB and PCI

The main characteristic of the technology is the ease of connecting and configuring peripherals to a host PC using the true Plug and Play concept without the need of opening the computer chassis. All add-on peripherals will be connected externally either through a direct USB port on the host or a USB hub unit (see Figures 1.3a and b). The USB has a 12 Mbps (full speed) and a 1.5 Mbps (low speed) specification to satisfy most low-end applications.

Definition of Universal Serial Bus (USB)

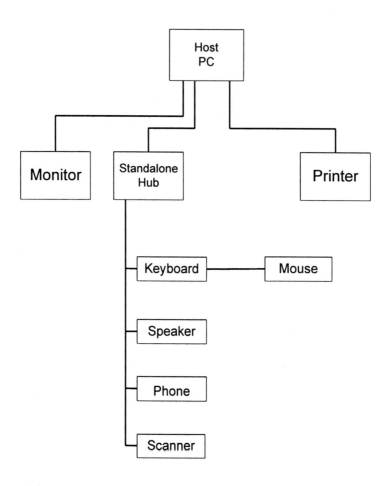

Figure 1.3a: USB Topology with Standalone Hub

Chapter 1: Introduction

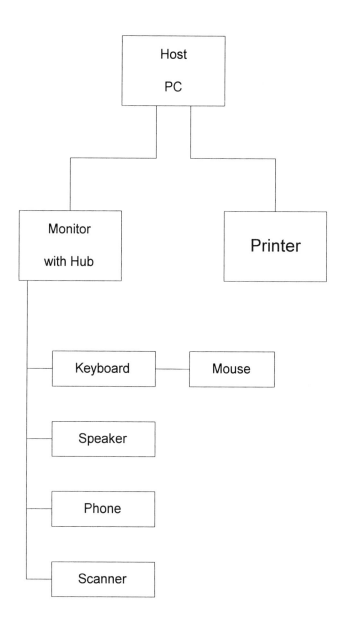

Figure 1.3b: USB Topology with Monitor Hub

1.3 Goals of USB

The USB specification 1.0 states the following goals for the USB technology:
- Ease of use for PC peripheral expansion
- Low-cost solution that supports transfer rates up to 12 Mbps
- Full support for the real-time data for voice, audio, and compressed video
- Protocol flexibility for mixed-mode isochronous data transfer and asynchronous messaging
- Integration in commodity device technology
- Comprehend various PC configurations and form factors
- Provide a standard interface capable of quick diffusion into product
- Enable a new class of devices that augment the PC capability

1.4 History of USB

The USB-IF has been developing a standard for the computer industry in the last few years. The current USB specification 1.0 is available to the public and can be downloaded from the web (www.usb.org). Hard copies are also available. Here is a summary of USB history.

1994 USB core companies formed.

1995 First WinHEC (Windows Hardware Engineering Conference) took place. USB-IF (USB Implementers Forum) formed. Membership was 340. Intel introduced first USB silicon.

Chapter 1: Introduction

1996 USB Specification 1.0 released. First Plugfest (compatibility workshop) took place. USB products introduced. First USB pavilion at Comdex.

1997 USB-IF membership increased to more than 400. Over 500 products were in development worldwide. First third party USB developers' conference held (sponsored by Annabooks). First USB participation at RetailVision.

1.5 USB Products Supported

There are many USB products supported today and more are being added to the list. The following is only a sample of product categories. (For updated product listings, see http://www.usbnews.com.) The strong momentum of USB technology will most likely make it a viable technology for the next decade.

- Host PC systems
- Printers
- Scanners
- Keyboards
- Mice
- Joysticks
- Gamepads
- Video cameras
- Still image cameras
- Telephones
- Modems
- Infrared devices
- ISDN adapters
- Mother boards
- Silicon
- Measurement Instrumentation
- Software and drivers

Software Cores
Diagnostic Systems
Wireless devices
USB cables and connectors

1.6 Market Potential and Opportunities for USB Peripheral Developers

According to Dataquest and Intel's projection (USB Conference July 96), USB PC shipments will be 20 million units in 1997 and 100 million units in 1999. Additionally, the Bishop Report stated that USB connector market would hit $400 million in 1999. The estimated ratio of peripheral use per host PC is four to one. Because of the ease of connection, USB home computer is expected to grow. Overall, USB peripheral product developers have tremendous market opportunities, which will continue to grow for the foreseeable future.

Chapter 1: Introduction

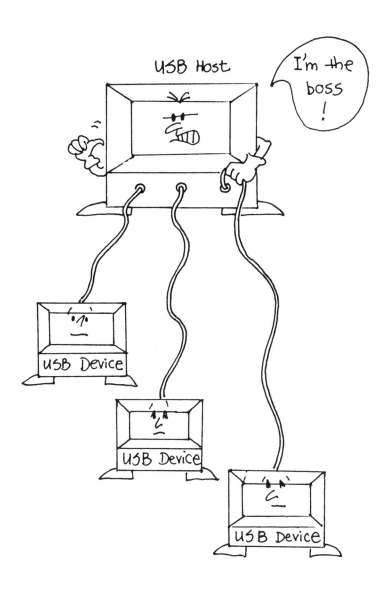

2. Organization of this Book

2.1 Purpose of this Chapter

The purpose of this chapter is to provide an overview of how the book is organized.

> Chapter 1: Introduction
>
> Chapter 2: Organization of this book
>
> Chapter 3: USB System Components Overview
>> This chapter provides a summary of USB features, overview of the USB systems, and review of the USB specification. Interlayer communication model and description of key USB components are presented from the viewpoint of a device designer.
>
> Chapter 4: Examples of USB Devices
>> This chapter provides examples of actual USB devices, including keyboards, joysticks, game pads, video cameras, infrared controllers, and so forth.
>
> Chapter 5: Designing USB Peripherals (Devices)
>> Provides design guidelines for USB Peripheral Developers as outlined in the following:
>>> Design Goals and Tradeoffs
>>>
>>> Working with USB System Components
>>>
>>> USB Device Components Selection (Chapter 6) Driver Issues
>>>
>>> USB Hubs Design Considerations
>>>
>>> USB Device Design Considerations
>>>
>>> Power Management Considerations
>>>
>>> Compatibility and Testing Issues

Purpose of this Chapter

> Cables and Connectors
> Other Issues:
> > FCC, EMI
> > UL, CSA, TUV
>
> Mechanical, Industrial Design And Prototyping

Chapter 6: USB Device Components Selection

> Provides information of over 30 Hub and Device ICs and selection criteria. Descriptions are included. Note that some of the silicon is only in the sampling stage. Check with the perspective vendors regarding availability.

Chapter 7: USB Development Tools & Help

> Provides a list of useful development tools. It includes both PC based products and stand alone units. For the most part, these tools have two key components: a signal generator, and a decoder and monitor to read the packets generated.
>
> This chapter also lists organizations which provide design and consulting services for hardware, software including drivers, silicon, mechanical, custom cables, custom connectors, seminars and training as a quick reference. A complete list of USB vendors is included in the appendix section.

Appendix: USB Power Application Notes, USB Power Distribution, Glossary, USB Frequently Asked Questions, and Vendor List.

Index

3. USB System Components Overview

3.1 Purpose of this Chapter

The purpose of this chapter is to describe the USB system and the role of each USB system component. As shown below, there are a total of five software and four hardware components to a USB system.

3.1.1 Key points in this chapter

Summary of USB Features
Back panel of the USB PC
Description of key USB System Components:
    Software Components:
        Windows 98 and USB Driver Development
        Client Software
        USB Driver
        USB Host Controller Driver
        USB Device Driver
    Hardware Components:
        Host Controller/Root Hub (see Interlayer model below)
        USB Hub
        USB Devices (see Interlayer model below) / Firmware
        Cable and Connectors
Review of the Interlayer Communication Model Applications

3.2 Summary of USB Features

As the name suggests, the Universal Serial Bus (USB will be used as an abbreviation from here on) is a serial bus topology, which communicates serially, and is intended to be used broadly in many applications. Do not confuse USB with RS-232. Though both are serial interfaces, USB is a bus topology and is much more powerful. See Figure 3.1 for a summary of the features.

Features	Description
12 Mbps (full speed) and 1.5 (low speed)	Two device speeds supported. Low speed is used for devices like mouse and keyboard. Unshielded cables can be used for low speed to reduce cost.
127 Devices	Up to 127 devices are supported on one USB bus.
Hot Plug	Devices can be attached to and detached from host PC without system power down. USB system software will detect the attachment and automatically configure the device.
Master / Slave scheme	The USB operates as a master / slave where the host PC controls everything. Don't be confused with a LAN or SCSI. USB does not allow sharing of printer by another host PC, for example.
Simple Cable / Connector Type	Only one type of cable (A to B connectors) is used for connecting devices to host PC. Two types of connectors are used to prevent loop back problem.
Bus / Self Power Choice	Two design options for the devices. Bus powered and self-powered are available to various applications.
System resources not required	Devices do not need system resources like IRQ and memory mapping.
Robustness	Error detection and re-send implemented.
Power saving	Suspend mode (sleep) reduce power to less then 0.5 mA per port.
Four communication models	Four types of Transfers are available for flexibility. They are Bulk, Control, Interrupt, and Isochronous.
Point- to- point connection	Devices can connect in any order. No external terminators needed.
Auto configuration	No jumper or IRQ setting necessary.

Figure 3.1: Summary of USB Features

Chapter 3: USB System Components Overview

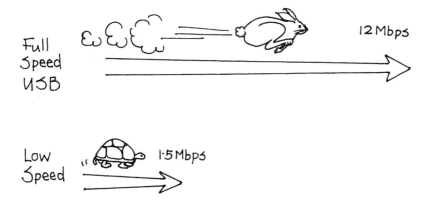

3.3 Back Panel of the USB PC

Figure 3.2 illustrates the back panel of today's USB PC. It consists of all the I/O adapters with two USB ports added. The future vision of the USB back panel simply has five connectors: two USB ports (some PC manufacturers are providing four ports) with one high speed IEEE 1394 (FireWire), one graphic and one LAN (I_2O) port (see Figure 3.3). The idea is to have all the main I/O ports built in the host PC so the users do not need to open up the PC chassis when adding peripherals. This future vision will take time to develop, and there are still a lot of open issues. For example, it is not clear whether the IEEE 1394 or the UltraSCSI port will take over the high-speed I/O port. However, Device Bay, a proposed method for allowing peripherals to be added after a PC is purchased, allows one USB port and one 1394 port.

Back Panel of the USB PC

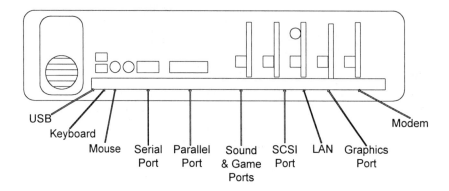

Figure 3.2: The Present USB Back Panel

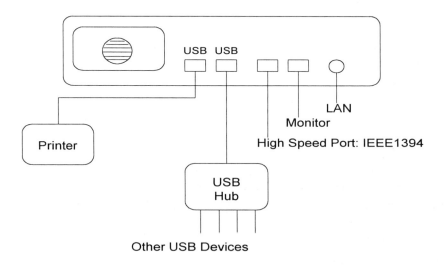

Figure 3.3: The Future USB PC Back Panel

3.4 Description of USB System Components

3.4.1 Software Components and Windows 98

Windows 98 and USB Driver Development

Microsoft Windows® 98 (previously code-named Memphis) is the next-generation operating system after Windows 95. It provides upgrades to Windows 3.x, Windows 95, OSR 2.0 and OSR 2.1. USB is among the many technologies supported. OSR2 will be further explained in a later section. The first Beta version, called Memphis Beta 1, was shipped at the end of Q197 with build number 1525. The Beta 2 version, officially called Windows 98, was shipped at the end of July 1997, with build number 1546. The Beta 2.1 version was shipped in early October 1997, with build number 1602.

To become a Beta tester for future releases of Windows 98, send email to memreq@microsoft.com.

For a list of key Windows 98 features and benefits, see the most recent *Microsoft Windows 98 Release Notes* (RELNOTES.DOC), distributed with each Beta release of Windows 98. The key Windows 98 features and benefits mentioned in the Beta 2.1 release notes include the following:

> Windows 98 can regularly test the hard disks, system files, and system configuration information to increase system reliability, and in many cases automatically fix problems. Enhancements to the Dial-Up Networking feature include the ability to link and synchronize multiple modems, and an ISDN Connection wizard makes it easier to configure hardware.
>
> The distributed component object model (DCOM) provides extensions for existing OLE interfaces.
>
> Enhanced FAT32 file system stores files more efficiently and frees up hard disk space.
>
> It is the fastest platform on which to run Microsoft Internet Explorer 4.0, which now supports Internet conferencing capabilities.

Description of USB System Components

Setup has been redesigned and streamlined.

System management improvements have been based on the policy-based central management guidelines and features outlined in the Microsoft Zero Administration Initiative for Windows (ZAW). For example, an Upgrade wizard provides smooth migration paths from Windows 95 and Windows 3.x-based systems.

Support for the newest generation of hardware, including cutting-edge media platforms, makes Windows 98-based systems easier to use, more entertaining, and more like everyday "appliances." For example, OnNow power management instantly starts a new computer, making it more like turning on a stereo or TV.

Support is included for the Intel MMX processor.

WDM (Win32 Driver Model) is a new unified driver model for Windows 98 and Windows NT 5.0, for target bus and device classes.

Universal Serial Bus (USB) support and IEEE-1394 (a.k.a. FireWire™) support enable more powerful device detection and the next generation of Plug and Play hardware.

Advanced Configuration and Power Interface (ACPI) supports easier device management on new computers and enhanced battery performance on new mobile computers.

Support for Digital Versatile Disk (DVD) and digital audio delivers high-quality digital movies and audio directly to a TV or computer monitor.

Support is provided in Windows 98 for the USB device classes listed below. Individual developers may provide other classes not listed. For the USB device class specifications, see http://www.usb.org; this includes the Common Class Specification, which defines device characteristics that apply across more than one device class (for example, power management at the device interface level).

Chapter 3: USB System Components Overview

Audio

Human Interface Device (HID), with native support for keyboard, mouse, and game controller devices

Hub

Imaging (for all the subclasses: moving image devices, supported by the WDM Stream Class driver, and still image and scanner devices, supported by the WDM Still Image Class driver).

Monitors

Universal Host Controller Interface (UHCI) and Open Host Controller Interface (OHCI)

The following USB device classes are not supported by WDM class drivers in Windows 98 Beta release 2.1. However, there are quite a few third party solutions available from vendors such as AMI, Elite, Ashley Laurent, SystemSoft, Award, and OS Group.

Printer

Communications

Mass Storage

Windows 95 and USB Driver Development

The initial release of Windows 95 (referred to as the Windows 95 Gold) has no USB support and is not upgradable to support USB. OSR 2.1 (OEM Service Release, 10/29/96), with the code name Detroit), is a supplement to the OSR2 (Windows 95 version 4.00.950b, Aug 96). It includes USB support. For more information about OSR 2.1, contact

http://www.microsoft.com/Windows/pr/win95psr.htm

Windows 95 provides USB support with OSR2.1, including the following features:

WDM at the 0.9 level.

UHCI and OHCI

USB Class Driver

Hub

Image Class Driver

Description of USB System Components

The OSR 2.1 Image Class Driver supports digital video cameras through the Image Class Driver, which is an interim solution for OSR 2.1 only. This driver is not part of Windows 98; support for digital video cameras in Windows 98 is provided by the WDM Stream Class. Support for digital still image cameras and scanners is provided in Windows 98 by the WDM Still Image Class.

In order for a host to support USB, it must have OSR2 installed (version 1111 or Win 4.00.950b). Additionally, it will need the OSR2.1. To verify OS2.1's existence on a platform, use the registry key (run REGEDIT.EXE to review):

HKEY_LOCAL_MACHINE \ SOFTWARE \ Microsoft \ Windows \ CurrentVersion \ VersionNumber

It should be 1212 or 1214.

To develop drivers for Windows 95 OSR 2.1, follow these steps:

Get the USB DDK from MSDN (it is part of the Windows NT DDK for Windows NT Workstation 4.0).

Install the USB DDK on your Windows NT 4.0 development platform, following the instructions in the readme file in the Windows NT 4.0 DDK subdirectory \win9x\usb.

For your target platform, use a PC with Windows 95 OSR 2.1 installed.

Windows NT 5.0 and USB Driver Development

Windows NT 5.0 release will support USB. For Windows NT 5.0 Beta release information, send email to nt5req@microsoft.com.

The process for developing drivers that run on Windows NT 5.0 platforms is similar to the driver development process for Windows 98; for more information, see the WDM DDK.

Note that Windows NT 4.0 cannot be upgraded to include USB support.

WDM Overview

To simplify hardware development, Microsoft introduced the Win32 Driver Model (WDM), which is a common set of I/O services and binary-compatible device drivers for both Windows NT and future 98 operating systems. This driver model is based on a class/minidriver structure that provides modular, extensive architectures for device support. It is a core technology for

> IEEE-1394 (Plug and Play)
>
> OnNow power management (Plug and Play)
>
> USB (Plug and Play)
>
> Zero Administration Windows (ZAW) initiatives

The WDM DDK (Device Driver Kit) provides the tool needed to develop functional drivers. The DDK is available through the Microsoft Developers Network (MSDN) and may also be distributed at USB Plugfests (see appendix section). Note that USB drivers built with the WDM DDK will work with either UHCI or OHCI. Both are host controller interfaces and talk to the host directly and are transparent to the device.

What Do You Need to Develop Drivers for USB Peripherals?

To develop drivers, you need to have an understanding of the USB Core Specification 1.0 (a new release of the Core specification is due in the first quarter of 1998), the USB device class specifications, and the operating system(s) for which you are developing the driver (Windows 98, Windows NT 5.0, and/or Windows 95 OSR 2.1).

The most current versions of the USB device class specifications are available from http://www.usb.org. The following lists the USB device class specifications, with their release status at the time of this publication. A 1.0 release is usable to design product and a 0.9 release is a complete description of the device class and ready for public comment.

1. Audio, 0.9
2. Communications, 0.9
3. Hub, 1.1

Description of USB System Components

4. HID (Human Interface), 1.0
5. Mass Storage, 0.9
6. Monitor, 1.0 release candidate
7. Power, 1.0 release candidate
8. Printer 1.0

At the time of this publication, the USB Image Class and Common Class specifications had not reached release 0.9 (they are still being worked on by the USB Device Working Group).

To develop USB device drivers for Windows 98 and Windows NT 5.0, follow these steps:

1. Get the WDM DDK from MSDN (or at a USB Plugfest or Developers' Conference).
2. For your development platform, use a PC with either Windows 98 or Windows NT 5.0 installed (from a developer's standpoint, the WDM DDK works on either Windows 98 or Windows NT 5.0).
3. For your target platforms, use a PC with Windows 98 installed and a PC with Windows NT 5.0 installed.
4. Helpful tools include Visual C++ version 5.0, or Visual C++ version 4.2 (version 4.0 has bugs) or other C compilers.

Finally, participate with both Microsoft testing and the USB Plugfests, sponsored by the USB Implementers Forum, to debug the drivers. Contact other vendors, including Intel, Toshiba, and other OEMs, to see if they are interested in testing USB devices in their laboratories.

Listing of Useful Web Sites

OnNow power management

http://www.microsoft.com/hwdev/onnow.htm

Public events for OEM and IHVs (Independent Hardware Vendors)

http://www.microsoft.com/hwdev/

Chapter 3: USB System Components Overview

Windows hardware quality lab

http://www.microsoft.com/hwtest/

Downloadable MS Word Viewer 97

http://www.microsoft.com/msword/internet/viewer

MS Word 6.0/95 Binary Converter

http://www.microsoft.com/Officefreestuff/word/dlpages/wrd6ex32.htm

UHCI specification

http://developer.intel.com/design/USB/designex/UHCI11D.HTM

USB/PnP white papers

http://www.microsoft.com/hwdev/busbios/usbpnp.htm

Bus enumeration code examples

http://developer.intel.com/design/usb/swsup/

http://developer.intel.com/design/usb/

Bus Design for PC Hardware

http://www.microsoft.com/hwdev/busbios/default.htm

Buttons for USB HID Devices

http://www.microsoft.com/hwdev/busbios/usbbuton.htm

Bus Specifications

http://www.microsoft.com/hwdev/specs/busspecs.htm

USB and Game Devices 2/13/97

http://www.microsoft.com/hwdev/devdes/usbgame.htm

WDM: HID Class Support 2/13/97

http://www.microsoft.com/hwdev/pcfuture/wdminput.htm

WDM: Press Releases

http://www.microsoft.com/corpinfo/press/1996/apr96/w32whcpr.htm

23

Description of USB System Components

Consultants and Training Seminars listing

http://www.microsoft.com/hwdev/wdmrsc.htm

http://www.annabooks.com

Vendors listing

http://www.usbnews.com

Client Software

The client software is the software resident on the host. It sends data to and receives data from a function and is independent of the hardware and USB. To illustrate how it works, let use choose a joystick design as an example. The client software in this case is the game software, which can be played by both USB and non-USB PCs. Another example of client software is an audio CD, which sends data to the function: the multimedia speakers mounted on the PC.

USB Class Driver (USBD)

This is a layer of software provided by Microsoft. It understands the communication requirements of handling an I/O request packet (IRP). Once a packet is received, it presents device information to the client software such as configuration and state management (endpoint, transfer type, transfer period, data size). It will either accept or reject the IRP based on whether the right procedure or protocol is followed.

USB Host Controller Driver (HCD)

This is an assistant to the client software, making the USB Class Driver hardware independent. Its tasks include the following:

    Scheduling of sending frames (one ms per frame typical).

    Tracking IRPs to and from a pipe.

    Initiating transactions via the root hub.

    Fetching data from memory and sending them out over the USB bus.

USB Device Driver (DD)

This is a layer of software responsible for initiating IRPs and communicating with the host. Because there are many kinds of Devices and, therefore, Device Drivers, "Device Class" (USB term) is formed to define the common characteristics of the device. For example, the HID (Human Interface Devices) defines the movement of a human hand when the HID, say a mouse, is used.

Currently, there are several classes being defined. They include the following:

>Audio Device Class
>Common Class
>Communications Device Class
>Hub Class
>Human Interface Device Class (HID) - keyboard, mouse, etc.
>Mass Storage Device Class
>Monitor Device Class
>Image Device Class
>Power Device Class
>Printer Device Class

3.4.2 Hardware Components

Host Controller/Root Hub

The Host Controller (HC) is responsible for generating transactions (scheduled by Host Controller Driver). The information includes the address of a device, transfer type, transfer direction, and address of the device driver's memory location. The SIE (Serial Interface Engine) is a transceiver responsible for converting parallel word to serial bits, handshaking, and error detection. Since all Hub and Device ICs have the SIE built-in, the above tasks are automatically taken care of.

Description of USB System Components

Intel and Microsoft have each defined their own host controller protocol. They both do the same thing with a different approach (See Figure 3.4).

While Intel implements UHCI (Universal Host Controller Interface), Microsoft, along with Compaq and National Semiconductor implement OHCI (Open Host Controller Interface). Others have joined in to support OHCI, including Symbios and Opti Semiconductor. Both UHCI and OCHI are supported by Windows 98.

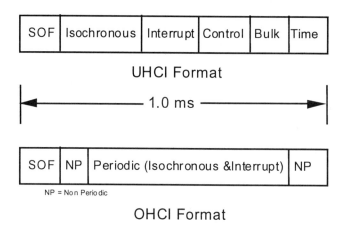

Figure 3.4: Host Controller Protocols

USB Hub

The USB hub provides a fan-out (from one to many) function for the host PC. Most of today's USB host PCs have two USB ports. With the hub it can be expanded to connect to many devices. Even though the USB specification supports a theoretical limit of 127 devices, in real life most users will have fewer than ten devices attached to a host PC.

The host PC is referred to as being upstream from the device while the device is downstream from the host. A USB hub is a device with one upstream and multiple downstream ports. For example, the USB HubSTAR Model 100 (made by Northstar

Chapter 3: USB System Components Overview

Systems) with one upstream port and four downstream ports is referred to as a four-port hub.

USB Devices

In the world of USB, a PC peripheral is referred to as a USB Device. It may consist of one or more endpoints. To relate this to the Interlayer Communication Model (ICM) block diagram (Figure 3.5), the device includes the Function (the peripheral itself), the USB logical device (firmware) and the USB Bus Interface (Hub or Device IC). In this case, the USB Device will include three blocks. Section 3.5 provides a more complete description of the Interlayer Communication Model. See Chapter 4 for examples of USB Devices.

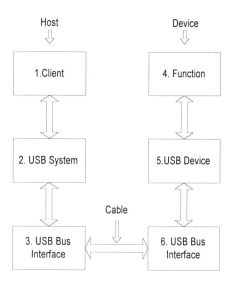

Figure 3.5: Interlayer Communication Model

Cable and Connectors

Today's serial and parallel port connectors include DIN, mini-DIN, DB-9, DB-15, DB-25, and more. All these will be replaced by the USB connectors, which have only two types: A and B. The USB

Description of USB System Components

cables will replace today's many serial and parallel cables, simplifying the PC-to-peripheral connection.

Figures 3.6a and 3.6b show the dimension of the overmold of the Series A and Series B cable plugs. A typical cable consists of a Series A to B plug (some manufacturers use the term receptacle instead of plug) at each end (Figure 3.6c). A detached or stand-alone cable (A to B) is used to connect a device to a Host PC. If the cable is attached to a device such as a mouse of joystick, only the A plug is used. A device developer will usually buy a "pig tail" cable from the cable manufacturers (Figure 3.8) with color to match the device. A typical USB cable (detached or stand-alone) is used to connect a Host PC to a device where the host (root hub) has a Series A connector and the device has a Series B connector. The Series A plug (on the cable side) has a rectangular shape and the Series B is more of a square shape. The drawings of series A and B connectors are shown in Figures 7a to 7e.

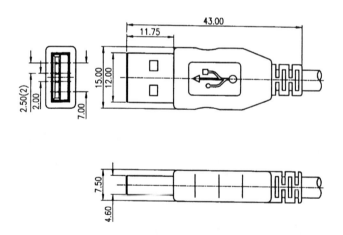

Figure 3.6a: Series A Cable Plug

Chapter 3: USB System Components Overview

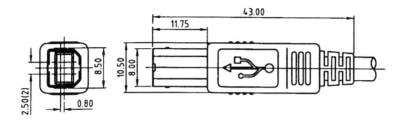

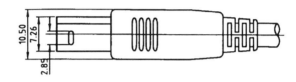

Figure 3.6b: Series B Cable Plug

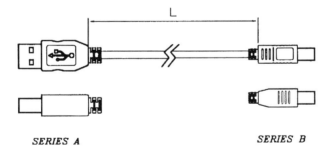

Figure 3.6c: USB A to B Cable

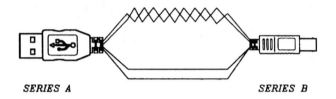

Figure 3.6d: Fully Rated Cable (shielding not shown)

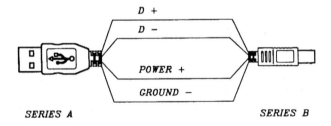

Figure 3.6e: Sub-Channel Cable (unshielded)

Chapter 3: USB System Components Overview

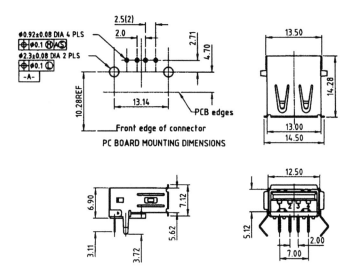

Figure 3.7a: Series A Receptacle

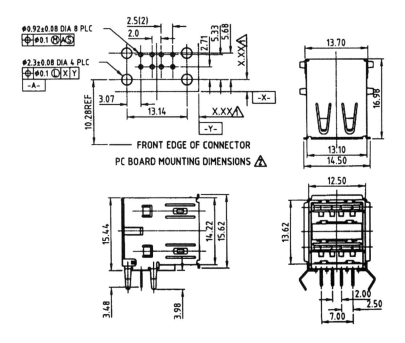

Figure 3.7b: Series A Receptacle (stacked)

Description of USB System Components

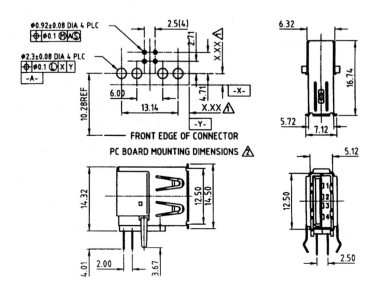

Figure 3.7c: Series A Receptacle (vertical)

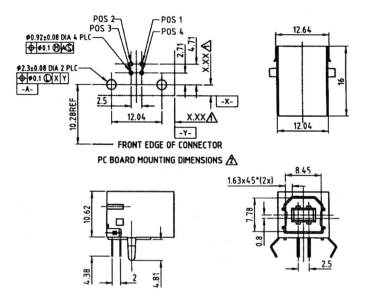

Figure 3.7d: Series B Receptacle

Chapter 3: USB System Components Overview

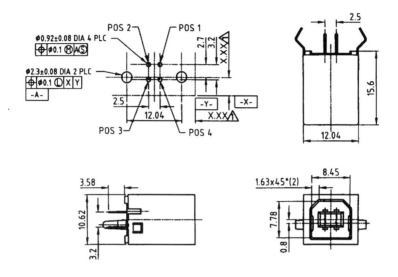

Figure 3.7e: Series B Receptacle (vertical)

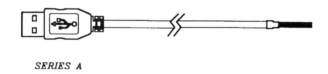

SERIES A

Figure 3.8: Pig-Tail Cable

There are two speeds supported by the USB: full (12 Mbps), and low (1.5 Mbps). To increase noise immunity of the cable for full speed operation, the shielded, twisted pair is used besides the differential signal requirements of the physical connection. The maximum length of the full speed cable is 5 meters and the impedance is 95 ohms +/- 15%. The low speed cable requires

Description of USB System Components

neither shielding nor twisted pair due to its low speed operation. Noise immunity is not as demanding. This will further reduce the cost of connection. The maximum cable length allowed is 3 meters. Figure 3.9 shows a summary of the USB cable characteristics.

Summary of USB cable characteristics

Material characteristics:

1. Jacket: PVC (polyvinyl chloride)
2. Color: Frost white recommended. Others are acceptable.

Cable characteristics:

1. Number of conductors: 4
2. Twisted pair requirement: one pair for fully rated channel (full speed) differential signal only. One twist per 6 to 8 cm with an overall jacket. None required for power and sub-channel (low speed) cables.
3. Gauge requirements: 28 AWG for signals, 20 to 28 AWG for power.
4. Jacket OD (outside diameter): 3.4 to 5.3 mm.
5. Color of conductors:

    + Data   Green
    - Data   White
    Vcc   Red
    Ground   Black

6. Shielding requirement: For fully rated cable only. Suggested shield is aluminized Mylar wrap with 28 AWG drain wire and 65% minimum coverage tinned copper mesh over foil.
7. Break strength: 45 Newtons minimum when tested in accordance with ASTM-D4565.
8. Maximum cable length: 5 meter (fully rated) and 3 meter (sub-channel)

Electrical characteristics:

1. Voltage rating: 30V (rms) maximum
2. Conductor resistance:

28 AWG 0.232 ohm/m
26 AWG 0.145 ohm/m
24 AWG 0.091 ohm/m
22 AWG 0.057 ohm/m
20 AWG 0.036 ohm/m

3. Fully rated only:

    Attenuation: (per ASTM-D-4566 not to exceed)

    0.064 MHz   4.8 dB/305 m
    0.256 MHz   6.7 dB/305 m
    0.512 MHz   8.2 dB/305 m
    0.772 MHz   9.4 dB/305 m

Chapter 3: USB System Components Overview

1.000 MHz	12 dB/305 m
4.000 MHz	24 dB/305 m
8.000 MHz	35 dB/305 m
10.00 MHz	38 dB/305 m
16.00 MHz	48 dB/305 m

Impedance: 90 ohm +/- 15% over 1 to 16 MHz. (ASTM-D-4566)
Propagation delay: <= 30 ns over the length of the cable

Environmental characteristics:

1. Temperature: -40 °C to 60 °C storage; 0 °C to 40 °C operating.
2. Safety approval: UL listed per UL Subject 444, Class 2, Type CM for Communication Cable requirements.
3. Flammability: Must meet NEC Article 800 or equivalent.
4. Marking: Vendor name, UL file number, type CM (UL).

Figure 3.9: Summary of USB Cable Characteristics

The VESA standard has recently defined the EVA (Enhanced Video Adapter), which includes the D+ and D- of the USB signals. This connector may gain popularity over time (see Figure 3.10).

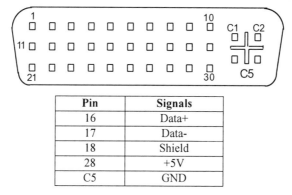

Pin	Signals
16	Data+
17	Data-
18	Shield
28	+5V
C5	GND

Figure 3.10: VESA Enhanced Video Adapter (EVA) Defines USB Pin-Outs

3.5 Review of the Interlayer Communication Model

The USB specification defines the interlayer communication model (Figure 3.5) between the host PC and a Device connected by a USB cable. On the Host side, there are three functional blocks: Client, USB System, and USB Bus Interface. On the Device side, there are three other blocks; Function, USB Device, and USB Bus Interface.

35

Each of these blocks consists of one or more of the system components described above.

3.5.1 On the Host side

Block 1-Client

This software sends data to or receives data from the function directly and expects certain tasks to be carried out. It does not care about how the communication works or how the packets are sent.

Block 3-the USB Bus Interface

This block is the Host Controller and SIE (Serial Interface Engine). In real life, the Host Controller is the motherboard with some USB chip set on it.

Block 2-USB System

This block includes software layers such as the USB Driver, Host Software, and Host Controller Driver included in the operating system. Memphis (Windows 98) and Detroit are the code names for such software, which include the USB drivers that talk to the Devices.

3.5.2 On the Device side

Block 4. Function (USB term)

This is the actual peripheral, such as a joystick.

Block 6. The USB interface block

This is the USB device IC (integrated circuit, silicon, or IC chip. The term IC will be used from now on.).

Block 5. The USB Device

This is the Device Driver that talks to the Host. The Device developers will have to provide this. Often, the device's silicon manufacturers supply sample code to make the device driver development easier.

3.6 USB System Operation Summary

A very simple way to explain the USB system is as follows:
- Step 1. The client software, such as a flight simulator game software, converses with the joystick directly. When the joystick is moved, the client software finds the right subroutine software and displays the turned scenery on the screen of the monitor. This will go back and forth. Note that the client does not care how many or how the packets are sent.
- Step 2. The drivers (however many there are) are busily taking orders from the client software, changing the instructions into packets. Their mission is to send these packets back and forth. These drivers have no clue what game is being played. As long as the packets are sent and received properly, the job is done.
- Step 3. Thanks to the Hub and Device ICs, the design engineers do not have to worry about converting the parallel words into serial bits, making sure the voltage level is correct and all packets received error free. All these tasks will be done automatically.

In short, the Device developers have to find the right Hub or Device IC, write the driver according to the specification, test the product (with other developers at the Plugfest, see Appendix), then ship the product. Simple!?

3.7 Application of the Model

When designing USB devices, there are three main concerns: Selecting the right USB Hub or device components (in some cases it may be designing the ASIC), making sure the device driver is

Application of the Model

available, and finally, observing all the I/O design rules. Chapter 5 will describe detailed design guidelines of USB devices.

4. Examples of USB Devices

4.1 Purpose of this Chapter

This chapter provides examples of USB Devices that were available when this book was being prepared:

Cables and Connectors from Northstar Systems, p. 40
Communications Interface from Ing Buro H. Doran, p. 43
Computer Attached Telephone from Mitel, p. 44
Digital Video Camera from Xirlink, p. 45
Gamepad from Alps, p. 45
Hand Input Device from Spacetec, p. 46
Host System from Toshiba, p. 47
IR Remote Control from Ing Buro H. Doran, p. 52
ISDN Controller from AVM, p. 53
Keyboard from Key Tronic, p. 55
Mother boards from Gateway 2000, p. 58
PC-to-PC Link from e^{TEK}, p. 59
PCI-to-USB Controller Board from CMD, p. 60
Printer Cable from In-System Design, p. 61
Remote Control from e^{TEK}, p. 62
Scanner from Logitech, p. 63
Speakers from Philips, p. 64
Stand-alone Hub from Northstar Systems, p. 66
USB-to-ECU Bridge from Ing Buro H. Doran, p. 67
USB-to-RS-232 Bridge from Ing Buro H. Doran, p. 68

The following information was received in response to inquiries sent to the USB community during the preparation of this book. The author apologizes to any vendors who may have submitted information that is not included here; material in incorrect format, lost material, or unreadable media may have been at fault. In any case, the author or publisher cannot be responsible for omissions, inaccuracies, or errors in the published information. The following information is intended as a guide to the reader as to the types and varieties of USB products becoming available.

4.2 Cables and Connectors

Company Name: Northstar Systems, Inc.

Product Name: USB Cables and Connectors
Features

Full line of Universal Serial Bus cables and connectors for ease of design:

Series A&B cables

    PC board mount single/stack series A receptacles

    PC board mount single series A receptacles (vertical)

    Series B receptacles

    Series B receptacles (vertical)

    Series A and B plugs.

Fully comply with USB specification 1.0 and UL/CSA requirements

Fully EMI/RFI shielded

Overmolding capability allows custom logo and part number insertion

ISO 9000 certified factories provides quality assurance

Description:

Northstar Systems, Inc. announces a family of Universal Serial Bus (USB) connectivity products. The product family includes the USB series A and B custom overmold cable assemblies, plugs and PC board mount receptacles.

The 4-conductor cable assemblies and connectors use phosphorous bronze as contact materials with selected gold plating at contact areas. The connector shells are made of steel. Two of four conductors are for power while the other two are for deferential signals. The power wire gauge accommodated ranges from 20 to 28 AWG (to supply power to downstream ports). The signal wire ranges from 26 to 28 AWG.

All cables comply with UL and CSA and exceed the USB specification version 1.0 speed requirement of 12 Mbps. Additionally, the cable assembly has been certified by the official USB-IF Plugfest workshop and guaranteed to be compatible with other USB-IF certified vendors. Custom logo overmold option is available. The cables come in a 1, 3, and 5-meter standard lengths. Custom lengths are available. Both series A and B plugs have similar specifications except for the dimensions. Series A is 12 mm x 4.6 mm x 11.75 mm (exposed area) and B is 8 x 7.26 x 11.75 mm (exposed area). Series A is for design with permanent cable attachment such as keyboard and mouse, while B is for detachable design such as the USB printers.

The PC board mount receptacles come in single and 2-in-1 stack versions. Though the USB implementer Forum suggested the single version as the standard design, system houses have already worked on designs using the stack version to provide additional USB ports at the PC back plane.

Northstar factories are ISO 9000 certified and comply with UL and CSA standards. Sales offices are located in USA, Ireland, Singapore and Taiwan.

Specifications

Materials

Insulators: UL 94V-0 rated polyester

Plug and Receptacle Contacts: 0.30 + 0.05 mm half-hard phosphor bronze.

Plating: Contacts are to be selectively plated with 1.25 micrometers (50 microinches) nickel over a minimum of 1.25 micrometers (50 microinches) copper over base materials.

Mating Area: Minimum 0.05 micrometers (2 microinches) gold over a minimum of 0.75 micrometers (30 microinches) palladium-nickel.

Solder: Minimum 0.38 micrometers (15 microinches) tin-lead over the underplate.

Plug and Receptacle Shell: Steel

Plug Outer Hood: PVC

Overmold color: Off-white, Gray, Charcoal, or Beige

Electrical

Contact Current Rating: 1 Amp per contact signal

Voltage Rating: 30 VAC (rms).

Mechanical

Wire Size Accommodated:

20-28 AWG (power)

26-28 AWG (signal)

Environmental

Temperature Rating: 0° C to 50° C

Company name:	Northstar Systems, Inc.
Address:	9400 Seventh Street, Bldg. A2,
	Rancho Cucamonga, CA 91730
Telephone:	909-483-9900
Fax:	909-944-0464
Email:	usbsales@northstar1.com
Web:	www.northstar1.com

4.3 Communications Interface

Company Name: Ing. Buro H. Doran

Product Name: USB CommSock

Description: This product is an interface between USB and the CommMan communications system. The CommMan is a stand-alone, eurocard format communications manager. It interfaces eight asynchronous serial ports, whose physical interface is determined by a piggy-back board known as a CommMod, with two synchronous/asynchronous high speed serial ports whose interface is also dertermined by a CommMod. Each CommMod services two serial ports. A socket on the motherboard also allows connection of a piggy-back module for higher speed communications such as Ethernet or CAN. This piggy-back module is known as a CommSock.

Using the 68PM302 controller from Motorola, with up to 2 MB Flash and 2 MB RAM on board, all driven by a highly efficient real time, multi tasking operating system with UNIX-like features, the CommMan can manage both standard and customer specific protocols on a port-by-port basis. GNU-GCC/GDB support will be available for the operating system in late 1997.

The USB CommSock is a high speed USB interface allowing the CommMan to act as a Communications Class device connecting eight standard serial interfaces with a host PC.

This product is for the industrial OEM who needs to connect different equipment such as measurement equipment or PLCs to a PC via the USB.

Company Name: Ing Buro H. Doran
Address: Beim Pfarrwaldle 7
              D-72813 St. Johann-Upfingen
              Germany
Telephone: +49 7122 82243
Fax: +49 7122 82263

Email: htd-ibhd@t-online.de

4.4 Computer Attached Telephone

Company name: Mitel

Product name: Mitel Personal Assistant

Description: The Mitel Personal Assistant is the industry's first telephony peripheral that supports the Universal Serial Bus (USB). With out-of-the-box Plug and Play, the Mitel Personal Assistant provides the same level of functionality as traditional plug-in card solutions without the need for time-consuming and technical system modifications.

Combining a small/home business's two most important tools - the telephone and the personal computer - the Mitel Personal Assistant simplifies messaging and communications management, providing home-based businesses both a competitive edge through a professional front-office image and a virtual administrative assistant that enables significantly increased productivity and improved customer service. Combined with the Plug and Play capabilities of USB, the Mitel Personal Assistant allows users to improve their most important information link to customers and colleagues: real-time voice communications.

The Mitel Personal Assistant, offering an integrated telephone/PC software solution, provides home-based workers new levels of ease-of-use and functionality. With the computer-attached telephone, users can - for example - place callers on hold with the click of a mouse button or use the PC to program special messages that are only played for important clients. Receiving an incoming call, the software quickly displays the caller's name, company, and telephone number. Simultaneously, users have quick access to notes previously made about the caller.

Company name: Mitel

Address: 350 Legget Drive

Kanata, Ontario

	Canada K2K 1X3
Telephone:	613-592-2122
Fax:	613-592-7825
Email:	alan_atkinson@mitel.com; peter_couse@mitel.com
Web site:	http://www.mitel.com/mpa

4.5 Digital Video Camera

Company Name:	Xirlink, Inc.
Product Name:	Video Phone

Description: The Xirlink Video Phone brings quality video conferencing to the home or office at an affordable price. Visionlink utilizes USB for compression and higher bandwidth and MMX technology for a smoother, faster frame rate. Visionlink ships complete with software, color digital camera, and USB cable and connectors for under $99.00. Visionlink is the perfect product for PC OEMs.

Company Name:	Xirlink, Inc.
Address:	2210 O'Toole Ave.
	San Jose, CA 95131
Telephone:	408-324-2100 ex: 446
Fax:	408-324-2101
Email:	dominic@xirlink.com
Web:	www.xirlink.com

4.6 Gamepad

Company Name:	Alps
Product Name:	USB Gamepad

Hand Input Device

Description: The Alps USB Gamepad is a USB game controller designed by professional game designers, producers, and testers from the consumer market. Designed for comfort and ease of use during extended play, this accessory offers players greater freedom and movement for fingers, so players can concentrate on gameplay instead of on the gamepad. The Alps Gamepad offers advanced controls in an easy-to-hold unit intended to meet the demanding needs of today's gamers.

1. Designed by the Pros - Gaming industry experts provided vital details that went into the development to make the unit user-friendly.

2. Ultra-smooth Direction Pad – Prevents thumb soreness after long periods of play.

3. Special Rubber Grip – The controller is designed with a rubber grip that offers players tactile feedback and a sure grip.

4. Hot Industrial Design - The gamepad's unique design was co-created by Smart Design, a New York-based developer of award-winning hand-held products.

5. Craftsmanship - Manufactured by Alps Electric, a 48-year-old company known for high-quality components and switches.

Company Name:	Alps Electric
Address:	3553 N. First Street
	San Jose, CA 95134-1804
Telephone:	408-432-6503
Fax:	408-321-8494
Email:	kkajikawa@alps.com
Web:	www.interactive.alps.com/usb

4.7 Hand Input Device

Company Name:	Spacetec IMC Corporation
Product Name:	SpaceOrb™ 360

Features:

Convenient hand-held design

Simultaneous six-axis (360 degree) control

10-bit digital precision

Rapid action buttons

Fully supported in Windows 95

User customizable

Easy to install

Description: The SpaceOrb 360 is a hand-held multi-axis (6 DOF) controller that allows you to intuitively perform any move you can imagine – even moves impossible with the keyboard, mouse, or joystick. Push, pull, or twist the SpaceOrb 360's Spaceball® PowerSensor® ball. Pitch, yaw, roll, strafe, jump, or cruise precisely in any direction, or combine them to perform difficult moves like Circle Strafes, Diving Rolls, SWAT Moves, and the indefensible Death Blossom.

The SpaceOrb 360 delivers true physics control by providing 10 bits on each axis of movement. You can instantly micro-adjust your speed, acceleration, the power of your punch, or how high you jump simply by varying the amount of force applied to the PowerSensor. Blast down hallways at lightspeed or slowly creep around corners. With the SpaceOrb 360's inuuitive digital control and a flick of the wrist, you can finally access the full speed range of the game, not just one or two speeds offered by the keyboard, mouse, and joystick.

Six rapid action buttons are at your fingertips for instant actions like firing, changing weapons, and opening doors. They're also fully programmable.

Windows 95 DirectInput™ support means that the SpaceOrb 360 works in any game designed for Windows 95.

The Customizer Utility personalizes the SpaceOrb 360 to your style of game play. Easily change button mappings, increase or decrease axis sensitivity, switch the SpaceOrb 360's orientation between horizontal and vertical, and choose from Introductory, Intermediate, and Advanced levels of game playing. Automatic Game Detection provides software that makes installation easy.

Company Name: Spacetec IMC Corporation

Address: The Boott Mills

Lowell, MA 01852

Telephone: 978-275-6100

Fax: 978-275-6200

Email: dperry@spacetec.com

Web: www.spacetec.com

4.8 Host Systems

Company Name: Toshiba

Product Name: Satellite 220CDS and Satellite 225CDS Notebook Computers

Features:

133MHz Intel Pentium processor

16MB EDO DRAM (expandable to 144MB)

256K L2 cache

12.1-inch 800 x 600 resolution DSTN color display

10X (average) CD-ROM drive

1.44 billion byte (=1.34GB) hard disk drive

Fast Infrared (FIR) port for fast file synchronization

Lithium Ion batteries

Two Type II (one Type III) PC Card slots, Card Bus ready and Zoomed Video ready

Description: The Satellite 220CDS continues the Toshiba tradition of providing customers with multimedia notebook computers at a competitive price. Toshiba's success in this customer segment is evidenced by the company's 50 percent market share in 1996 for portable computers below $1,999, according to Audits & Surveys.

Product Name: Satellite Pro 440 and 445 series
Notebook Computers

Features:

800x600 resolution displays, TFT, or the new FastScan display (12.1 inches)

133MHz Intel Pentium processor with MMX technology

PCI bus

16MB of EDO DRAM (expandable to 144MB)

256K L2 cache

1.44 billion byte (=1.34GB) hard disk

Slimline SelectBay version of a 10X CD-ROM drive and floppy disk drive are also included standard

33.6 Kbps voice/data/fax/modem

Two Type II (one Type III) PC Card slots

CardBus

Zoomed Video Ready

Description: The latest in display technology and MMX processing meet in the new Satellite Pro 440 Series. The new feature-rich systems target a mid-$2,000 price point where one of every three notebook computers sold is a Toshiba system, according to Audits and Surveys 1996 data. The Satellite Pro 440 Series is designed to continue this trend by providing the best available technology in this price class.

Host Systems

Product Name: Satelllite Pro 460CDT
Notebook Computer

Features:

166MHz Pentium processor with MMX technology

256K Level 2 cache

32MB RAM (expandable to 160MB)

2.02GB hard disk drive

32-bit PCI bus architecture

Integrated 33.6Kbps, Cellular Ready V.34 data/fax modem with full duplex speakerphone and answering machine capabilities

12.1-inch TFT active matrix color display with 800 x 600 resolution and 16 million colors

Software MPEG or hardware MPEG support through a Zoomed Video Card

Two Type II (one Type III) Card Bus ready PC Card slots

Interchangeable SelectBay versions of a 10X CD-ROM and floppy disk drive

Description: The most powerful member of the Satellite Pro family, the new Satellite Pro 460CDT is an exceptional price/performance offering. With Intel's fastest mobile processor -- the 166MHz Pentium processor with MMX technology -- 256K Level 2 cache and 32MB RAM (expandable to 160MB), the newest Satellite Pro system is capable of handing the demands of sophisticated business applications.

Product Name: Port,g, 300CT
Notebook Computer

Features:

133MHz Intel Pentium processor with MMX technology

256K L2 cache

PCI architecture

32MB of DRAM standard

1.51GB hard drive

Advanced panoramic 1024 x 600 resolution active matrix color display technology

Fast Infrared (FIR) port allowing for information to be transmitted at 4Mb per second to other infrared equipped peripherals

Two Type II (one Type III) PC Card slots that support Zoomed Video (ZV) and CardBus are included for easy expandability

Integrated 33.6Kbps voice/data/fax modem

16-bit Sound Blaster stereo sound

Description: At only 3.8 pounds and 1.4-inches thick, the new Port,g, 300CT re-establishes a true ultra-portable form factor without compromising performance, setting a new standard for this class of machine. The Port,g, 300CT is perfect for the mobile professional seeking more power and productivity without being forced to carry a six-to eight-pound traditional notebook.

Product Name: Tecra 520CDT and Tecra 530CDT

　　　　　　　　　　Notebook Computers

Features:

166MHz Intel Pentium processor with MMX technology

256K pipeline burst SRAM L2 cache

32MB EDO DRAM expandable to 160MB

2.02GB removable hard drive

Built-in 33.6Kbps data/fax/voice cellular ready modem

10X (average) CD-ROM drive

PC Card functionality using Zoomed Video (ZV) and CardBus Technology

12.1-inch 1024 x 768 high-resolution TFT active matrix color display with 64K colors

HiQVideo PCI video controller, 2MB EDO video memory, 64-bit graphics acceleration

Description: Designed for corporate users, the new Tecra notebook computers pack a powerful combination of performance, multimedia, and expansion in a new, lighter-weight industrial design. The new Tecras offer professionals the ideal mobile computing platform to support more flexible work patterns - whether working at home, working at multiple sites or working in a conference room with a team - with the most value for their dollar.

Company Name:	Toshiba Computer Systems Division
Address:	P.O. Box 19724
	Irvine, CA 92618
Telephone:	800-457-7777
Web:	http://computers.toshiba.com

4.9 IR Remote Control

Company Name:	Ing Buro H. Doran
Product Name:	IR2USB

Description: This product is an IR remote control transmitter / receiver unit for OEMs and end users. It is an addition to a family of IR remote control units offered by this company. The product range includes decoders for PC and 19" rack systems with RS-232 and PC-keyboard outputs. Handsets can be customer-specified at little or no extra cost. Programmed chips, either transmitter or receiver, for design-in applications are also available.

The IR2USB module is a small module providing a low speed USB interface to a host PC. The product can currently be delivered as either a pre-programmed microcontroller chip with design-in information or as a finished unit with indicator LEDs. Available as options are relay outputs for applications requiring both the sending of commands to a host system and direct control of equipment without interference from the host.

Modular software allows the IR2USB to be programmed with a customer-specific IR protocol, allowing seamless integration into present equipment.

Company Name: Ing Buro H. Doran

Address: Beim Pfarrwaldr 7 D-72813 St. John

Telephone: +49 (0) 7122 82243

Fax: +49 (0) 7122 82263

Email: htd-ibhd@t-online.de

4.10 ISDN Controller

Company Name: AVM

Product Name: ISDN-Controller B1 USB

Features:

Active ISDN Controller for the basic rate interface with high-speed RISC CPU and 1MB memory

Description: The AVM ISDN-Controller B1 USB is member of the most powerful and popular family of active ISDN-Controllers on the market today. The B1 USB offers a stable platform for professional users, who require performance, power and reliability.

The AVM ISDN-Controller B1 USB is the link between the ISDN network and your personal computer. This USB controller for personal computers with USB interface handles simultaneously the two B-channels (2 x 64 Kbit/s) used for data transmission and the D-channel (1 x 16 Kbit/s) reserved for the monitoring and execution of protocol functions. The AVM ISDN-Controller B1 USB may both be connected to directly to the ISDN network or an ISDN private branch exchange (PBX).

ISDN Controller

The B1 USB supports the following D-channel protocols: X75, X25 (ISO 8208), X.31, V.110, V.120, transparent, T90 (Fax G4), T30 (Fax G3), ISO3309 (for GSM).

For international use, AVM offers several D-channel protocols, such as E-DSS1, 5ESS, NI1, CT1, VN3, 1TR6, and Austel (Australia).

The ISDN USB controller is completely software controlled. The controller has a flash BIOS that makes updating the on-board software very simple.

Currently, AVM provides COMMON-ISDN-API 2.0 drivers for Windows 95 and, in the future, will provide for other Plug-and-Play enabled operating systems.

The B1 USB has a flexible hardware architecture. The complete ISDN-protocol software stack is executed "on board" the B1 USB controller, via a downloadable file. Additionally AVM delivers drivers for the Microsoft ISDN-Accelerator Pack (NDIS WAN for Windows NT and 95 and the CAPI Port Driver for Windows 95). These drivers are true CAPI 2.0 applications and establish the connection between Microsoft's network services and the ISDN USB controller.

AVM offers a broad range of applications for this controller, including FRITZ!32. It includes file transfer, fax G3 application (up to 14400 baud sending on both B-channels), fax journal, terminal emulation, and an answering machine. Other applications, especially for networking purposes like routers or remote node applications, are also available.

Performance Summary:

Plug-and-Play USB controller for personal computers

USB interface and controller software for Windows 95

S0-interface for ISDN basic subscriber access

Designed for high performance operation with two B-channels

Simultaneous processing of both B-channels using the full transfer rate

High-performance microprocessor (a 20-MIPS transputer) and 1 MB RAM on board

Data transfer rate 2 x 64 Kbit/s und 1 x 16 Kbit/s

Applications interface COMMON-ISDN-API 2.0

Company Name:	AVM Computersysteme Vertriebs GmbH & Co. KG
Address:	Alt-Moabit 95
	D-10559 Berlin, Germany
Telephone:	030-399 76-185 / -0
Fax:	030-399 76-299
Email:	a.ziessnitz@avm.de / u.knack@avm.de
Web:	www.avm.de

4.11 Keyboard

Company Name:	Key Tronic Corporation
Product Name:	Key Tronic KT2000 -USB
Model:	E03640USxxx-C

Features:

104 key layout with standard enter key and Windows 95 GUI and application keys

Data output serial 1.5Mb/s USB protocol

Standard type A USB connector

Attractive enclosure design, overall size 8.28" wide, 18.41" long, and 1.89" height at back edge

Switch life of 30 million cycles

Low cost, high quality membrane technology. Dynamically attachable and configurable

Keyboard

Supports concurrent operation of many devices on a single peripheral bus

Description: This new keyboard adds Plug and Play to the KT2000 line of keyboards. It provides USB developers unprecedented flexibility and ease of use for attaching a keyboard peripheral. Dynamic configuration allows insertion and removal at any time, even when the host system is up and fully operational. The keyboard is a self-identifying peripheral and configures upon attachment as defined in the Rev 1.0 USB specification.

All of the information required for setup and operation of the keyboard is stored in the keyboard ROM in segments called descriptors. At time of attachment to an operating host, (hot plug) the keyboard sets up and enables endpoint zero for communication to begin. Through this default control pipe the keyboard receives commands and requests for data from the host operating system or BIOS. Data for device, configuration, interface, endpoint, and string descriptors is then available to the host via standard USB requests.

The host reads the device descriptor to determine the maximum packet length needed to talk to the newly connected device. The device descriptor also contains information to identify the manufacturer, device model number, the USB specification to which the product is designed, device revision, and other global attributes particular to this device. Next, the host will instruct the keyboard to change its address to a non-zero address. Address zero is reserved for all devices to default to when they are first connected or powered-on. The actual address is determined by the topology of the bus at the time the device is attached, and may be different every time the keyboard is connected.

A USB operating system running on the host then reads the configuration descriptor. The configuration descriptor includes attributes of the keyboard and how it expects to operate, including, but not limited to, power consumption, whether or not the device is bus powered or self-powered, endpoint characteristics, report content and length for the data sent by the device, etc. Some devices may support more than one configuration. Model E03640USxxx-C supports only one configuration, but the host must still send it a set configuration request before it will operate.

After configuration, endpoint 1 operates as an interface endpoint and the keyboard is ready to communicate with the host system. A keystroke will now cause the keyboard to send a packet of information to the host at the next available poll of the endpoint. If no changes in key status (key press or release) occur between consecutive polls of the interrupt endpoint the keyboard responds with a NAK, indicating nothing has changed.

The keyboard supports the set-idle request. The idle rate determines how often to send a report of the keyboard's current key status, regardless of whether or not any keys have changed since the start of the idle period. Idle can be disabled or it can be set by the host to any period between 4 milliseconds and 1.02 seconds with a 4 millisecond resolution. When idle is disabled the keyboard will report the status of the keys on the keyboard each time endpoint 1 is polled. The default idle rate established at power-up is set to report every 500 milliseconds. The host can request real-time status of the keys without affecting idle by using the get-report request and reading the status of the keys back through the control pipe.

Since a keyboard is the most common boot device for a personal computer, model E03640USxxx-C supports boot protocol. At power up, the BIOS interrogates devices connected to the bus looking for boot devices. When it identifies the keyboard as a boot device, the BIOS assigns an address and sets the configuration to one. The BIOS will then issue a set-protocol request to enter boot protocol. Boot protocol is a special predefined operating mode that permits the BIOS to read and understand keystrokes without having to read and understand all the descriptors. This allows the keyboard to be used during system configuration for the selection of options, features, etc. After power-up, system operation is passed from the BIOS to the operating system, which will read the descriptors and load appropriate drivers to control the keyboard. The keyboard will then operate in the report protocol as described by the descriptors. Support of both protocols is critical for system configuration, real mode DOS applications, and general use.

Model E03640 complies with HID specification Version 1.0 Draft #4 for human input devices. As a low speed device it transfers data

Motherboards

at a rate of at 1.5 megabits per second. This allows for access to the bus without incurring excessive expense in EMI shielding, and widens the frequency tolerance, which permits the use of a lower cost resonator instead of a crystal as the device clock oscillator source.

Company Name: Key Tronic Corporation

Address: P.O. Box 14687 Spokane WA 99214-0678

Telephone: (509) 927-5520

Fax: (509) 927-5503

Web: www.keytronic.com

4.12 Motherboards

Company Name: Gateway 2000

Product Name: Portland

Model Number: Klamath 266 FPC/Klamath 266 XL

Description: Gateway 2000 system based on Intel 440FX and PIIX3 chipset

Product Name: Mailman II

Model Number: P55c-200FPC/P55c-166FPC

Description: Gateway 2000 system based on Intel 430VX and PIIX3

Product Name: Venus

Model Number: G6-200 Professional/G6-200FPC

Description: Gateway 2000 system based on Intel 440FX and PIIX3 chipset

Product Name: Lawman

Chapter 4: Examples of USB Devices

Model Number: P55c-200FPC/P55c-166FPC/P55c-133FPC

Description: Gateway 2000 system based on Intel 430TX and PIIX4 chipset

Company Name: Gateway 2000

Address: 610 Gateway Drive

North Sioux City, SD 57049

Telephone: (605) 232-2000

Fax: (605) 232-2781

Email: stepptho@gw2k.com

4.13 PC-to-PC Link

Company Name: e^{TEK} Labs

Product Name: USB Lynx ™

Features:

Compliant with USB Specification 1.0

Works with Microsoft Windows 95 OEM Service Release 2.1

Compatible with NDIS 4.0

Plug and Play ease of installation and use

Data transfer rates approaching 10 Mb/s

Description: The Lynx provides PC-to-PC connectivity via the Universal Serial Bus. The link between the PCs appears to be a network link to the user. This allows the user to use all of the built-in Windows 95 networking functions; no proprietary software is required.

Company Name: e^{TEK} Labs

Address: 1057 East Henrietta Rd.

Rochester, NY 14623

Telephone: 716 292-6400

Fax: 716 292-6273

Email: Ptravers@eteklabs.com

Web: www.eTekLabs.Com

4.14 PCI-to-USB Controller Board

Company Name: CMD Technology

Product Name: PCI-to-USB Controller Board

Model Number: CSA-6700

Features:

Quick, reliable, and low-cost solution for connecting current and future product offerings to the USB

Upgradability for existing systems which do not have USB. Computer OEMs can add in the card and ship Microsoft's OHCI-based USB drivers.

Higher performance than other USB solutions

Complete software support. CMD offers a solution with full software support needed by peripheral OEMs to sell their devices to the retail market.

Description: The CSA-6700 is the industry's first standalone PCI-to-USB controller board that is based upon CMD's own USB0670 100-pin ASIC, the industry's first standalone PCI-to-USB controller chip. The board contains dual USB connectors, and connects up to 127 devices. The CSA-6700 is compatible with the OpenHCI specification. Drivers supported include DOS, Windows 95, and Windows 3.x.

Company Name: CMD Technology, Inc.

Address: 1 Vanderbilt

                    Irvine, CA 92618

Telephone: 800-426-3832

Chapter 4: Examples of USB Devices

	714-454-0800
Fax:	714-455-1656
Email:	info@cmd.com
Web:	www.cmd.com

4.15 Printer Cable

Company Name: In-System Design
Product Name: Printer Cable with Instant USB™
Features:

USB connector and one-meter shielded USB cable on one end

Centronics "B" connector on the other end

VLSI logic (USS-720) integrated into the Centronics connector does protocol and electrical level conversion between USB and 1284 Parallel Port specifications

Compatible with USB hub devices

Compatible with all parallel printer devices

Transparently negotiates the highest level of 1284 the printer supports (compatibility, nibble, EPP, and ECP modes)

Compatible with Microsoft USB Windows 95 Printer Class driver

Fully Plug and Play compatible

Can print from any application running under Win 95, including DOS apps

Dissipates approximately 300 milliwatts from the USB bus

Description: This product is a "Smart Cable" that, through hardware and software, provides intelligent and transparent protocol conversion between Universal Serial Bus (USB) on the PC side and IEEE-1284 Parallel on the printer side. It will allow any Windows PC with a USB port to print to any parallel printer. The heart of device is the USS-720 Instant USB ™ chip from Lucent Technologies Microelectronics Group.

Remote Control Device

The cable connects USB-equipped portable or desktop PCs running Windows 9x with any parallel printer. Windows NT 5.0 support will be available when 5.0 releases.

Company Name: In-System Design

Address: 12426 West Explorer Drive, Suite 100

Boise, ID 83713

Telephone: 208-377-9222

4.16 Remote Control Device

Company Name: e^{TEK} Labs

Product Name: KwiKey IR ™

Features:

Infrared remote control for easy access to PC functions

Utilizes standard television-style remote control

Completely programmable via the KwiKey configuration software

Executes keystrokes and mouse movements for system commands and applications

Replaces complex keystroke and mouse movements with a simple button press

Low cost solution for OEMs. Licensing available

Perfect for bundling into PCs or for integration into PC monitors or consoles.

Available in both wireless and custom-wired versions.

Description: The KwiKey IR Remote Control is a highly advanced remote control for your computer. It can have from 5 to 30 buttons or more; each programmed for a specific function. You no longer need to wade through windows or find specific applications to perform needed tasks. You can just hit a button on the KwiKey IR

Remote Control. If your needs change, you can easily reprogram the buttons for different functions.

The KwiKey Remote Control is available in both wired and infrared versions. The infrared version has a receiver that plugs into the back of your computer. You are then free to wander around the room while you deliver a business presentation, or sit back in your favorite chair while you surf the Web. Both versions provide true Plug and Play functionality, and offer dynamic insertion and removal (hot-swapping), allowing you to plug in the device and immediately use it without rebooting your computer.

Company Name:	e^{TEK} Labs
Address:	1057 East Henrietta Rd.
	Rochester, NY 14623
Telephone:	716 292-6400
Fax:	716 292-6273
Email:	Ptravers@eteklabs.com
Web:	www.eteklabs.com

4.17 Scanner

Company Name:	Logitech
Product Name:	PageScan USB

Features:

True USB connectivity - hot Plug and Play

No power supply required - powered up by USB port and one cable less to drag around the table

Multifunctionality - provides scan, copy, fax, text, and image editing in one compact tool

Compact design - fits anywhere on a busy desktop

24 bit color, 300 X 300 optical resolution and 2400 X 2400 enhanced resolution

New intuitive user interface - ScanBank - enhances users' scanning experience

Description: Logitech PageScan USB is a compact, full-page color scanner that takes advantage of USB technology to provide easy installation without turning off your system or even rebooting. 24-bit color scanning with 300 dpi optical resolution and 2400 dpi enhanced resolution. Scan photos, artwork, and text with great results. Send faxes with your fax modem and make copies with your printer. Innovative Logitech ScanBank software integrates Xerox TextBridge OCR and gives you a natural way to store, organize, and use scanned items. Includes Adobe PhotoDeluxe. Logitech's advanced engineering eliminates the need for an additional power supply.

Company Name:	Logitech
Address:	6505 Kaiser Dr.
	Fremont, CA 94555
Telephone:	510-795-8500
Fax:	510-796-8058
Web:	www.Logitech.com

4.18 Speakers

Company Name:	Philips Semiconductor
Product Name:	DSS 220 and DSS 350 Multimedia Speakers

Features:

High sound quality

Interference-free

Better acoustics

Hot Plug and Play

Does not require soundcard

Extended featuring possibilities

Extended flexibility

(for DSS 220 only:)

  Bass reflex speaker

  USB Master-Slave system

Connections:

  Digital in (USB)

  Analog in

  Subwoofer out

  Headphone out

  Local digital volume control

  Dynamic bass enhancement

  Auto standby detection

  USB controls   Volume

                          Bass

                          Treble

Description: Philips Electronics, technology leader in USB silicon and peripherals, provided test products, including speakers, to the Windows 98 Beta sites. Philips Sound Systems is now ready to introduce the first product from a range of USB speakers for the desktop.

These speakers are aimed at the upgrade market with outstanding design and excellent price performance ratio. The feature set of the DSS 350 is richer. These speakers allow both digital and analog inputs. This makes them very attractive and future proof even for people not having the latest technology available on PC yet. These products have been demonstrated at major PC and CE shows around the globe.

These products have been developed in parallel with OS development. They are working with all the audio support currently available in Microsoft's OS release called " Windows 98 ".

Technical Specifications:

Power:	10 Watt system
Freq. Response:	60 – 20.000 Hz
Input Voltage:	12 Volt
Mains supply:	120 Volt
Dimensions:	9.6" x 5.4" x 4.8"
Weight:	1.5 Kg / set
Company Name:	Philips
Address:	1070 Arastradero Rd.
	Palo Alto, CA 94304
Telephone:	415-846-4408
Fax:	415-846-4466
Email:	acheil@pmc.philips.com

4.19 Stand-alone Hub

Company:	Northstar Systems, Inc.
Product Name:	USB HubSTAR™

Features

   Compact size with 4 or 6 downstream ports

   Self-powered capability to provide 500 mA to downstream port

   Individual port over-current protection

   Complies with USB specification 1.0

   ISO certified factories for quality assurance

Description: The USB HubSTARTM hub allows four USB devices to be connected to the host PC. A total of five tiers of hubs will allow the host PC to connect up to 127 devices. The automatic self and bus-power sensing technology enables the user to simply plug in the AC adapter to make the unit self-powered with 500 mA for each downstream port. If the device from downstream has a circuit malfunction and causes the port to be shorted to ground, the built-in circuit protection will automatically shut down the port from damages. When the unit is in the suspense state, the current consumption drawn from upstream is less than 500 nA. This compact unit is stackable and easy to use.

Specifications:

Number of ports:

One upstream port with detached 2-meter series A to B cable

4 downstream port version (Model: USB HubSTAR 100) or 6 downstream port (Model USB HubSTAR 200) are available

Dimension of unit:

1.0 inch high x 6.5 inch wide x 3.75 inch deep

Power: Self-powered or bus-powered

Certification: UL and CSA certified with FCC class B approval.

Company name:	Northstar Systems, Inc.
Address:	9400 Seventh Street, Bldg. A2
	Rancho Cucamonga, CA 91730
Telephone:	909-483-9900
Fax:	909-944-0464
Email:	usbsales@northstar1.com
Web:	www.northstar1.com

4.20 USB-to-ECU Bridge

Company Name:	Ing Buro H. Doran
Product Name:	USB2ECU

Description: This product bridges the USB and an ECU (ISO9141) interface. USB2ECU is a special adaptation of the high speed USB2RS232 interface offered by this company. It allows the connection of an ECU to a PC. Thanks to on board intelligence, the task of managing bandwidth-consuming ECU protocols is offloaded to the interface. The interface can optionally be supplied with flash memory to enable on-line software updates or programming by the customer. Status LED's allow visual indication of ECU bus traffic.

The USB2ECU is fully ISO9141 compliant and allows the connection to 12V or 24V automobile systems.

ECU protocols offered are KW2000 and KW71. Other protocols can be added as desired.

Company Name: Ing Buro H. Doran

Address: Beim Pfarrwaldr 7 D-72813 St. John

Telephone: +49 (0) 7122 82243

Fax: +49 (0) 7122 82263

Email: htd-ibhd@t-online.de

4.21 USB-to-RS232 Bridge

Company Name: Ing Buro H. Doran

Product Name: USB2RS232

Description: This product bridges the USB and an external RS-232 interface. USB2RS232 is another member of the low to mid level protocol converter family offered by this company. The system is an intelligent bridge between a high speed USB and a serial interface with RS-232 signal levels. The unit can optionally be delivered with flash memory to enable the in-circuit programming of customer-specific serial protocols, thus offloading bandwidth-consuming protocol servicing to the interface.

A special option allows the connection, via piggy-back modules, of standard interfaces such as TTY, RS422/RS485, and customer-specific modules.

This product is for OEMs needing to integrate their PCs in an industrial environment.

Company Name:	Ing Buro H. Doran
Address:	Beim Pfarrwaldr 7 D-72813 St. John
Telephone:	+49 (0) 7122 82243
Fax:	+49 (0) 7122 82263
Email:	htd-ibhd@t-online.de

5. Designing USB Hubs and Devices

5.1 Purpose of this Chapter

Designing a marketable product is always easier said than done. In this chapter, we will attempt to provide some guidelines to help you avoid pitfalls and speed up your design process. Design considerations of bus-powered hubs, self-powered hubs, bus-powered devices, and self-powered devices will be provided. Information on hub, device, and power ICs from the leading suppliers is available in Chapter 6.

5.1.1 Key points in this chapter

- Design Goals And Trade-offs
- Driver Issues
- USB Hubs Design Considerations
- USB Device Design Considerations
- Power Management Considerations
- Compatibility And Testing Issues
- Cables And Connectors
- Other Issues:
    - FCC, EMI
    - UL, CSA, TUV
    - Mechanical, Industrial Design And Prototyping Custom ASIC

5.2 Design Goals and Trade-offs

There are five design goals that a designer will constantly try to achieve. Keep in mind that a successful product is one that has the optimal combination.

1. Best performance
2. Lowest cost of development
3. Lowest product cost.
4. Highest product quality
5. Shortest time to market

5.3 Driver Issues

In Chapter 3, the tasks of a DD (USB device driver) were explained. A DD is needed for the host to work with the Device. There are four sources of DDs:

1. Use the generic DD available in Microsoft's Windows 98.
2. Write your own DD.
3. Buy the DD and firmware.
4. Hire a consultant to develop custom DD.

5.3.1 Generic Device Drivers Available in Microsoft's Windows 98

The intent of the operating system (OS) Windows 98 (to be released as Windows 98) is to include all the possible generic drivers so that when a device is attached, the OS will automatically recognize the device. Windows 98 includes some drivers like the generic Hub Drivers, which have been tested in the Plugfest. Other DDs, like the Audio Class, are still being defined. You may need to provide a special device driver to take advantage of special features not supported by the generic device. This is sometimes referred to as a "mini DD" used to supplement the Windows 98's generic DD. Microsoft is very eager to incorporate new drivers or new features to its list.

5.3.2 Write Your Own DD

If the decision is to write your own, be sure to obtain all the required guidelines to make the driver work with the OS (Windows, NT, etc.). The information is available from the following sources.

- WDM (Win32 Device Module) guideline from Microsoft is the blueprint for writing the working device driver. It is included in Windows 98, which is available to qualified beta sites or USB developers with an NDA (non-disclosure agreement) signed.

- WDM (as part of DDK, Device Developer Kit) is also available to MSDN (Microsoft Developer Network) members and requires a membership fee.

Other sources of information and training include:

- Annual WinHEC (Window Hardware Engineers Conference) sponsored by Microsoft. They provide WDM workshops.

- Training and Seminars available from Annabooks, TeKnowledge, and others.

- WDM technical papers available from the Microsoft Web site.

5.3.3 Buy the DD and Firmware

If one decides to take this shortcut, some software suites are available from SystemSoft (Printer, Monitor, and Serial to Parallel Adapter). The advantage of this approach is that the software suite on the host side has been tested and it guarantees that the device will talk to the host.

5.3.4 Custom DD

Similar to Choice 2, a consultant with experience in writing Window drivers will save you valuable time (see the Appendix for some vendor listings).

5.4 USB Hub Design Considerations

USB Specification 1.0 has defined specific USB Hub requirements. Figure 5.1 illustrates a simple block diagram of the basic USB Hub functions. The repeater handles all the activities relating to connectivity and the controller handles communication with the host. Fortunately almost all of the hub functionality is included in the features of many commercial hub ICs. The designer needs to be able to select the best hub IC for the design.

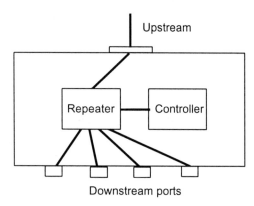

Figure 5.1: Hub Function

5.4.1 Basic Hub Functionality

Basic hub functionality includes the following:

1. Port connect/disconnect. Activities include hub configuration, port state management, detection of connect/disconnect of downstream port devices, and suspend/resume. The specific port states are power off, disconnected, enabled, disabled, and suspended.
2. Bus fault and reconfiguration.
3. Power Management (See power section below).
4. Full- and low-speed support.

Chapter 5: Designing USB Hubs and Devices

The USB specification further defines a new class called Compound Device, which is a combination of the hub and another function in one single unit. An example will be a keyboard hub, comprising a hub and a keyboard in one unit.

5.4.2 Hub Feature Considerations

USB hubs can be stand-alone located in a monitor, keyboard, or even a printer. The initial product offering from hub vendors will most likely be a stand-alone or monitor hub. In hub design, many features need to be considered:

1. Number of downstream ports - 3, 4, 6, 7, or 8 with optional embedded port. If it is a monitor hub, an Inter Integrated Circuit (I^2C) will most likely be used for signal connection.
2. Speed - must support 12 and 1.5 Mbps.
3. Power - bus or self-powered. In addition, when designing power switching, there is a choice of per port, 2-port, or ganged power switching.
4. Capable of configuration per USB Specification 1.0 (see USB Specification 1.0 compliance below).
5. LED indicators (optional, see power section).
6. FCC class A is a minimum requirement. Class B is preferred (see below for explanation).
7. UL CSA, and CE certification is a minimum requirement.
8. Upstream cable - detached or attached for bus-powered hub. Attached for self-powered hub.

5.4.3 Hub IC Component Selection Criteria

Chapter 6 outlines the description of various Hub and Device ICs for your reference. There are 3, 4, 6, 7, or 8 ports available from the Hub ICs. Additionally, some provide an optional embedded port. The minimum requirement for a hub IC is to be able to perform the tasks in Chapters 9 and 11 of the specification.

Specification 1.0 compliance

All ICs will promise USB Specification 1.0 compliance in their product literature. Indeed, this is a minimum requirement. This means that the IC will be able to support full/low-speed, suspend/resume, hub config-uration, handshaking, and error detection. It also has a built-in SIE/transceiver to convert the parallel word to serial bit stream.

Since there is an update to Chapter 11 (hub spec. chapter), make sure that the IC will meet the new requirement as well. The update is available from USB-IF at no cost. Additionally, the USB Plugfest (compatibility workshop) allows USB-IF member vendors to test interoperation among devices.

Integrated microcontroller

As seen in Figure 5-1, there is a repeater and a controller in the hub block diagram. When selecting the hub IC, it is important to know that some IC vendors offer the repeater portion while others offer both the controller (microcontroller) and the repeater in an integrated IC.

The number of downstream ports is three or four. Some vendors offer a six, seven, or eight port version. If you are designing a monitor hub, check to see if the I^2C feature is available. It is an embedded port, which provides signal connection to a monitor.

Power - bus or self-powered - is built-in for some ICs. It is a matter of connecting the self-power pin high or low. This can be a very convenient feature.

Noise immunity capability can be a tricky thing to handle. If the IC provides a better filter at the transceiver level, it may mean your hub can pass FCC class B easier.

Throughput is probably the ultimate measurement of performance for some applications. Even though the rating of full-speed is 12 Mbps, a USB system, including a hub, seldom achieves this theoretical limit. The state machine, the software, and other overhead may cause the aggregate throughput (actual transmission rate of useful data after the

IC overhead is deducted) to be between 30% to 90% of the theoretical limit. It will not be an issue if a low-speed device, such as a mouse, is connected. How-ever, if multiple isochronous sources are sending packets through one upstream port of the hub, it may make a difference. Most IC vendors do not provide throughput performance data. The only way to measure it is to use a traffic generator to create bit stream of data through the upstream port to determine how many packets are being transferred.

5.5 USB Device Design Considerations

Compared with the hub, the device design is simpler. Design considerations include the following:

- Speed - Full- or low-speed. Note that a 1.5K pull-up resistor will need to be tied to D+ (3.0V to 3.6V) to indicate that it is a full-speed device when attaching to the USB bus (Figures 5.2a and b).

- Power – High- or low-power (see power section below).

- Capable of configuration per USB Specification 1.0 (mostly done by the Hub ICs).

- FCC class A is a minimum requirement. Class B is preferred. (See below for explanation.)

- UL, CSA, and CE certification is a minimum requirement. (See below for explanation.)

- Device cable - detached or attached.

USB Device Design Considerations

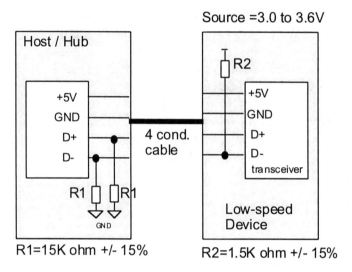

Figure 5.2a: Low-Speed Device Cable and Resistor Connection

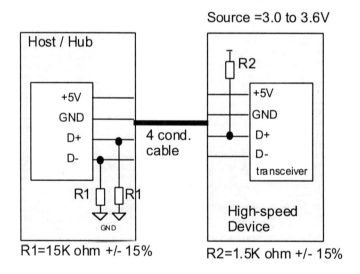

Figure 5.2b: High-Speed Device Cable and Resistor Connection

5.6 Device IC Components Selection Criteria

See Chapter 6 for description of various Hub and Device ICs for your reference.

Specification 1.0 compliance

All ICs will promise USB Specification 1.0 compliance in their product literature. Indeed, this is a minimum requirement. This means that the IC will be able to support full- or low-speed device configuration, handshaking, and error detection. It also has a built in SIE/transceiver to perform the parallel word to serial bit stream. It should pass the USB Plugfest (compatibility workshop) that allows USB-IF member vendors to test interoperation among devices.

Integrated microcontroller should be included.

Built in FIFO buffer ranging from 64 to 512 bytes.

Built in clock.

Noise immunity capability can be a tricky thing to handle. If the IC provides better filtering at the transceiver level, it may mean your device can pass FCC class B easier.

Power Considerations

Bus-powered Hub

High-powered Device

5.7 Power Considerations

The USB specification 1.0 provides design rules of power management for bus or self-powered hub, high- and low-power devices.

Here is a summary of the USB power management requirements:

Chapter 5: Designing USB Hubs and Devices

Overcurrent protection is 5 Amp per port for self-power devices for safety reasons. If a faulty downstream port draws more than 5 Amp, the downstream port will be removed from the power source. Proper operation is guaranteed with initialization.

Voltage drop is 4.75V for self-powered ports and 4.4V for bus-powered ports. All devices should operate between 3.3V and 4.4V.

During configuration, no device will draw more than 100 mA. After configuration, if power is available, the device will draw the current allowed. If the power is not available, it will be reported to the user.

A high-power device will draw a maximum of 500 mA and a low-power device will draw a maximum of 100 mA.

A high-power port will provide a maximum of 500 mA and a low-power port will provide a maximum of 100 mA.

In suspend mode a device should not consume more than 500uA.

5.7.1 Power Management Design considerations

Self-Powered Hub

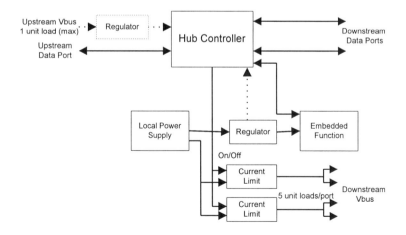

Figure 5.3: Compound Self-Powered Hub

Power Considerations

Unlike a bus-powered hub, a self-powered hub can have more than 4 downstream ports.

A self-powered hub has an advantage over the bus-powered one in that a local power source will provide multiple LED indicators with the needed current. Because a hub has to support both full and low speed, a 1.5K pull-up resistor to D+ is needed to indicate that it is a full-speed device.

The hub can have overcurrent protection and reporting on a per-port, two-port, or a ganged basis. It is a trade-off between features and cost.

A self-powered hub can be stand alone, part of a monitor, a printer, or a keyboard. For the stand-alone hub, a power source such as a wall transformer will be needed. Depending on the brand, it may be cheaper to have the regulation of 5V built in the wall mount power adapter.

Bus-Powered Hub

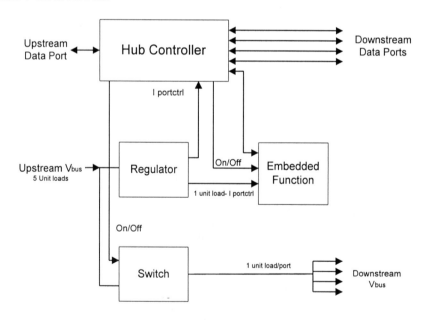

Figure 5.4: Compound Bus-Powered Hub

Chapter 5: Designing USB Hubs and Devices

A bus-powered hub can have one upstream port and four downstream ports.

When connected to a 500 mA upstream port, the hub will draw current from upstream and supply each downstream port with a maximum of one unit load or 100 mA. The controller and optional embedded ports combined cannot draw more than 100 mA.

When connected to a 100 mA upstream port, only the controller and the embedded port will be powered, assuming that the two ports will not draw more than 100 mA. The other downstream ports will not have power.

A 1.5K pull-up resistor is needed to indicate a full-speed device.

Self-Powered Devices

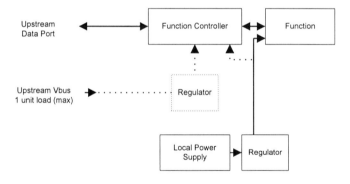

Figure 5.5: Self-Powered Function

Most likely, it is a full-speed device. If it is, a 1.5K pull-up resistor will need to be connected to D+.

The function controller can draw power from either the upstream or the local power source. If the controller is powered from an upstream port, the detection and enumeration will take place automatically even though the local source is turned off. On the other hand, if the controller is powered by a local power source and the power is turned off, no detection will take place.

83

Power Considerations

Once configured, the device can draw a maximum of 100 mA from an upstream port.

Bus-Powered Devices

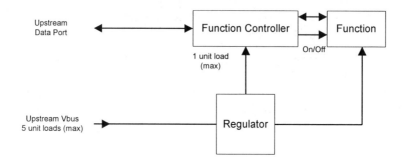

Figure 5.6: High-Power, Bus-Powered Function

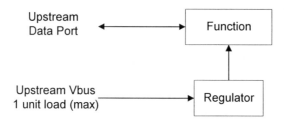

Figure 5.7: Low-Power, Bus-Powered Function

The design of low-powered device is simpler. Since either a 100 mA or 500 mA port will provide sufficient power, there are no special design considerations.

If a high-power device is connected to a 500 mA upstream port, it will start to work after configuration. On the other hand, if it is connected to a 100 mA upstream port, sufficient power will not be available for the device and the condition will be reported to the user.

5.8 Compatibility and Testing Issues

One of the challenges of developing a new USB device is to make sure that it is compatible with other USB devices. Offered free to USB-IF members is the Plugfest, a test workshop held bi-monthly for developers to test their USB products among themselves. USB-IF will certify those products after they have passed Chapter 9 of the USB specification. Make sure the devices are tested with both the UHCI and OHCI.

An easy way to test the devices more vigorously is to have multiple hubs and devices stacked 5-tier deep (preferably using multiple brands) and loop Chapter 9 tests for at least 24 hours. By using this method, some of the intermittent problems may be discovered that might be otherwise undetected (See Chapter 7 for available development tools).

5.8.1 Cables and Connectors

Cables are easily overlooked in the design process. Frequently, it is not addressed until the very last moment. This can be costly if it prevents a launching of a major product. Consider the following when designing/selecting a cable and a connector:

> Custom cable - Most of the custom cable houses will provide a way to insert a custom logo on the cable itself. Lead-time is a frequently overlooked. Allow at least four weeks for the development of a new custom logo insert after receipt of artwork. Additionally, when choosing a custom color, specify it with a PMS color number. Avoid general terms such as light gray or dark gray. Allow additional lead-time for custom color.

> Noise problem- Problems can be generated by many factors. One of them is lack of proper grounding of circuitry and poor shielding of cables. Make sure cables for full-speed are specified as shielded twisted pair for the differential signals. If noise is caused by poor system grounding and it is not feasible to redesign the system because of the deadline,

85

consider using ferrite core on the outside of the cables. See appendix.

Custom length - For volume purchases of 5,000 or more, most custom houses are willing to produce cables with custom length (no more than 5 meter for full-speed and 3 meter for low-speed per USB spec.). Once again, custom length will add four weeks to the process time, especially when dealing with overseas factories.

5.8.2 Other Issues

FCC, EMI

Compliance with FCC class A specifications for industrial use is required by law to ship products in the USA. FCC class B is more difficult to pass than class A but is required for home use.

If the hub case is plastic, it may be necessary to have an inner metallic coating for increased noise immunity. For shielded cable, make sure that there is proper shielding design (joints with no gaps); otherwise it will defeat the purpose of shielding.

When designing a compound self powered hub, make sure that the upstream and downstream ports have a common ground. Their bus voltage (V_{BUS}) should be isolated. Additionally, the chassis and signal ground should be DC isolated. Otherwise, there may be a problem with the low frequency current present.

UL, CSA, CE

UL (USA) and CSA (Canada) are agencies who deal with safety due to potential fire hazard. Fire retardant plastic and cable jacket is needed. Choose UL, CSA approved factories. CE is a similar agency for Europe. Seminars are available to teach people how to achieve compliance. (See vendor listing.)

5.8.3 Mechanical, Industrial Design, and Prototyping

To shorten the time to market, consider using new plastic molding techniques which are capable of producing plastic case samples in days instead of weeks.

5.9 Custom Application Specific Integrated Circuits (ASIC)

Finally, for those who prefer developing custom hub or device ICs there are three alternatives sources.

>Write your own codes.
>
>Buy the core logic. There are vendors who have developed standard core available for licensing. Usually, a one-time fee is charged for a specific application such as a hub. See chapter 7 for a list of core products.
>
>Develop custom codes. Hire a consultant to write the code and develop the IC. See Chapter 7.

6. Examples of USB Hub and Device ICs

6.1 Purpose of this Chapter

This chapter provides examples of USB Hub, Device, and Power ICs that were available when this book was being written:

Device ICs

   Alcor Micro, p. 91

   Anchor Chips, p. 92

   CMD, p. 96

   Cypress Semiconductor, p. 99

   Future Technology Devices, p. 102

   Lucent, p. 106

   Mitsubishi, p. 111

   Motorola, p. 113

   Nogatech, p. 115

   Samsung Semiconductor, p. 116

   ScanLogic, p. 118

   USAR Systems, p. 120

   Winbond, p. 121

Host ICs

   Cypress Semiconductor, p. 121

   Intel, p. 122

Purpose of this Chapter

Hub ICs
    Alcor Micro, p. 130
    Atmel, p. 132
    Cypress Semiconductor, p. 134
    Future Technology Devices, p. 134
    KC Technology, Inc., p. 136
    Lucent, p. 141
    MultiVideo Labs, p. 145
    National Semiconductor, p. 146
    NetChip, p. 148
    Philips, p. 151
    Texas Instruments, p. 153
    Thesys, p. 157
    Winbond, p. 159

Power ICs
    Unitrode, p. 159
    Micrel, p. 162

Other ICs
    Analog Tranceiver from Philips, p. 163
    Audio Converter from Philips, p. 164
    Scanner Control Interface from Winbond, p. 165
    USB-to-Parallel Conversion IC from Lucent, p. 166

The following information was received in response to inquiries sent to the USB community during the preparation of this book. The author apologizes to any vendors who may have submitted information that is not included here; material in incorrect format, lost material, or unreadable media may have been at fault. In any case, the author or publisher cannot be responsible for omissions, inaccuracies, or errors in the published information. The following information is intended as a guide to the reader as to the types and varieties of USB products becoming available. Please also see the list of vendors in Chapter 8.

6.2 Device ICs

Company Name: Alcor Micro, Inc.
Product Name: USB Device Controller
Model Number: AU9201
Features:

 Single chip USB device controller

 Compliant with USB specification 1.0

 Configurable to full and low speed

 Power management including suspend and resume

 Fault detection and recovery

 Integrated USB transceiver

 Serial bus interface engine (SIE) for packet decoding and generation

 CRC generation and checking

 NRZI encoding/decoding and bit stuffing

 Built-in pair of transmit and receive FIFOs

 Input pin for overcurrent sensing

 Output pin for port power switching

 Microcontroller-based architecture for easy customization for derivative products

 Integrated customized device firmware (e.g. USB mouse)

Device ICs

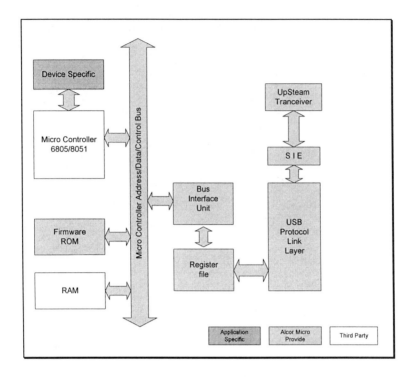

Company Name: Alcor Micro Inc.
Address: 155A Moffet Park Dr.
Suite 240
Sunnyvale, CA 94089
Telephone: 408-541-9700
Fax: 408-541-0378
Email: Dennis@Alcormicro.com

Company Name: Anchor Chips
Product Name: USB Device Controllers
Model Number: AN2131Q and AN2151Q
Features:

Single-chip, low-power solution for high-speed USB peripherals

Tiny footprint

Complete solution in 44 PQFP

USB Specification 1.0 compliant

Thirty-one flexible endpoints

Supports isochronous, bulk, control, and interrupt data

"Soft" configuration management

8 or 32 Kbytes of memory

Low-power 3.3v operation

Five times speed of standard 8051

Supports composite devices

I^2C controller

Multiple enumeration without disconnect

On-board PLL

Description: The Anchor Chips EZ-USB™ family provides significant improvements over previous USB architectures with the inclusion of an enhanced 8051 core and 8 Kbytes or 32Kbytes of RAM. The enhanced 8051 core provides five times the performance improvement over the standard 8051, while maintaining complete 8051 software compatibility. With the multitude of 8051 development tools, designers can stay within their current design environment. With 8 Kbytes or 32 Kbytes of RAM, the 8051 operating code and data memory can reside inside the chip, eliminating the need for external memory. The 8 Kbyte version, AN2131Q, comes in two packaging configurations; a 44 PQFP and an 80 PQFP. The 32 Kbyte version is pin- and software-compatible with the AN2131Q. Both versions have ROM equivalents to allow easy migration for code protection.

The AN2131Q contains an enhanced 8051 core, 8 Kbyte dual-port RAM, I^2C bus, USB Core Engine, three I/O ports, two UARTs, USB transceiver, timers, and on-board PLL. It conforms to the high-speed (12 Mbps) requirements of USB Specification 1.0, including support for remote wake-up. The internal SRAM replaces EPROM, OTP EPROM, or masked ROM function that is conventionally used in other USB solutions, providing maximum flexibility. The pin-compatible AN2151Q has the same features as the AN2131Q with 12 Kbytes of RAM for larger program needs. For customers who need to protect their firmware code, a masked ROM option, which is pin- and software-compatible to the RAM version, is available.

The focus of this chip family is to provide a flexible design solution using a RAM architecture for "soft" configuration while still providing a choice of design options for customers. The soft configuration enables peripheral manufacturers to accommodate code changes due to field updates, last minute software code changes prior to production, or dynamic changes in peripheral properties as set by the user. There are four options for loading 8051 firmware to generate USB traffic:

1. *Loading from a software file.* This provides the maximum flexibility to the peripheral manufacturer. This configuration takes advantage of the internal 8K or 32K of RAM to load 8051 boot code as provided by the host system. Because of the proprietary Renumeration™ capability of EZ-USB chips, a new set of descriptors can be loaded after the initial enumeration without physically disconnecting the device. This allows device descriptors and 8051 program code to be loaded from a customer-supplied driver disk. Only the Vendor ID and the Product ID need to be loaded during boot time in hardware through a 16 byte EEPROM. Using this configuration, users can implement the USB function in a tiny 44 PQFP package yielding a complete USB solution in less than one square inch of a PC board.

2. *EEPROM loaded through I^2C port.* This gives designers the capability to load an entire 8051 program code of 8 Kbytes or less from hardware. Because of the flexibility of the external EEPROM and internal 8K RAM, manufacturers have the option to make last minute changes to their design or code without impacting the production schedule.

3. *External memory loaded through the memory expansion port.* If the internal RAM space is not enough for program memory,

program code can be loaded through the memory expansion port. With the 80 PQFP package, separate 16 bit address and 8 bit data ports are available to directly attach to a 64K EPROM, SRAM, or flash ROM that has direct access to the enhanced 8051 core. Unlike a standard 8051, the address and memory ports are not multiplexed, eliminating the need for glue logic for connection to external memory.

4. *Internal ROM for peripheral manufacturers who migrate to the ROM-based AN2133Q.* The pin- and software- compatible option, AN2133Q, allows customers to load 8051 firmware from an internal ROM source. This provides the ability to protect firmware code while still providing RAM-based devices for development and initial production.

The enhanced 8051 processor core has been designed to offer increased performance by executing instructions in four clock cycles as opposed to twelve clock cycles in the standard 8051. This shortened bus timing improves the execution rate for most instructions by a factor of five.

Feature	Standard 8051	Enhanced 8051 Core
Clocks per instruction cycle	12	4
Data pointers	1	2
Serial ports	1	2
16 bit timers	1	3
Interrupt sources (int. and ext.)	5	13
Stretch memory cycles	No	Yes
Nominal operating frequency	12 MHz	24 MHz
Nominal operating voltage	5 v	3.3 v

Company Name: Anchor Chips
Address: 12396 World Trade Drive, M/S 212
San Diego, CA 92128
Telephone: 619-613-7900

Device ICs

Fax: 619-676-6896
Email: sales@anchorchips.com
Web: www.anchorchips.com

Company Name: CMD Technology
Product Name: PCI-to-USB Controller Chip
Model Number: USB 0670
Features:

Quick, reliable, and low-cost solution for connecting to the USB.

Higher-performance than other USB solutions

Because the USB0670 is based upon the OpenHCI specification, it is very efficient as far as processor overhead and bus bandwidth are concerned.

Complete software support

Description: The USB0670 100-pin ASIC is the industry's first standalone PCI to USB controller chip. The product is ideal for X86, Pentium, and Pentium-Pro based desktops; portable computers; docking stations, RISC-based platforms, and other PC workstations; and other computing devices. A 3.3 Volt version (USB0673) is available in a TQFP package.

The USB0670 is based upon the OpenHCI specification and is fully compliant with the PC97 Hardware Design Guide.

Company Name: CMD Technology, Inc.
Address: 1 Vanderbilt
Irvine, CA 92618
Telephone: 800-426-3832
714-454-0800
Fax: 714-455-1656

Email: info@cmd.com
Web: www.cmd.com

Company Name: CMD Technology
Product Name: Keyboard Controller
Model Number: USB0678KMp

Features:

Designed to work seamlessly with all UHCI and OHCI-based USB host controllers, including CMD's own PCI0670 PCI to USB host ASIC.

Patent pending PS/2 mouse daisy-chaining capability allows the use of commodity PS/2 mice

USB keyboards based upon the USB678KMp are fully supported by major BIOS companies, ensuring operation with all operating systems and during POST

Full support by Microsoft's HID driver for Windows

Low cost 42-pin DIP package

Integration of power-on reset circuitry and other features cuts down on external components further reducing the total solution cost.

Description: The USB0678KMp Universal Serial Bus keyboard controller ASIC is 100% compliant with the latest USB specifications, and provides a low-cost intelligent design that allows keyboard manufacturers to reliably upgrade from the old PS/2 style keyboards to the advanced capabilities provided by USB. In addition, the USB0678KMp provides extended features and customization capabilities that will add differentiation and a competitive edge over more generic USB solutions

Because the USB0678KMp replaces the existing PS/2 keyboard controller, all the benefits of USB are available at a very low incremental cost. In addition, the USB0678KMp has patent pending "virtual hub" capabilities, allowing commodity PS/2 mice to be daisy chained from the keyboard. This enables personal computer manufacturers to get full USB keyboard and mouse functionality with the lowest system cost configuration possible. With the addition of our Hub controller, the keyboard can become a USB console into which other USB devices such as joysticks or gamepads may be connected.

CMD's membership on the USB committees assures that the USB0678KMp was designed in 100% compliance with the industry standard, allowing low cost keyboards to operate reliably with all industry standard host controllers. Use of the low cost subchannel of USB means that expensive cables and EMI suppression components necessary for full speed operation on USB may be eliminated. For lower cost applications, special ROM masks can be made for high volume custom configurations.

The USB0678KMp has a downstream PS/2 mouse port allowing PS/2 mice to be connected to the keyboard. The USB0678KMp then converts the PS/2 information into USB protocol, resulting in "virtual hub" capability.

Microsoft will provide the Human Input Device (HID) Driver to OEMs that request it for Windows 95 and NT. In addition, with BIOS support from all BIOS companies, CMD's USB keyboards function with all operating systems and applications, including those that do not support USB natively

CMD's USB Partners Program

CMD makes integrating USB into your keyboard design a quick and painless process. They provide sample chips, drivers, and PCI USB controller boards that incorporate efficient testing and integration into your latest designs. Because keyboards often operate in noisy environments, CMD's USB Partners Program provides critical time-to-market differentiation with proven reference circuitry and layout guidelines, resulting in quick reliable designs.

CMD also makes available to qualified CMD USB Technical Partners their custom firmware and driver design service. CMD's extensive software and firmware utilities maximize the

Chapter 6: Examples of USB Hub and Device ICs

performance and reliability of USB keyboards. Partners can also choose to design their own firmware or drivers, and CMD will provide certification in our extensive testing facility. CMD also provides, free of charge, as part of their USB Technical Partners Program, a 24-hour USB Bulletin Board Service, allowing users to integrate the latest USB host drivers and updates into their products.

Company Name: CMD Technology, Inc.

Address: 1 Vanderbilt

Irvine, CA 92618

Telephone: 800-426-3832

714-454-0800

Fax: 714-455-1656

Email: info@cmd.com

Web: www.cmd.com

Company Name: Cypress Semiconductor

Product Name: Low- and High-Speed Microcontrollers

Model Numbers: CY7C63xxx, CY7C65xxx, CY7C66xxx, CY365x

Features:

Integrated 8-bit RISC microcontroller core

Industry's smallest RISC core

Optimized USB instruction set (34 instructions)

On-chip clock multiplication

Integrated 128/256 bytes SRAM

Integrated 2/4/6/8 KB EPROM

USB Serial Interface Engine (SIE)

USB transceiver

10-40 General purpose I/Os

Full power management

Description: The Cypress USB microcontrollers are a full family of low- and high-speed solutions for USB peripherals. The family is built on the industry's smallest RISC core providing the most highly integrated solution on the market.

The Cypress USB solutions are self-contained units developed using an 8-bit RISC-based microcontroller core with integrated EPROM, SRAM, full-custom USB serial interface engine, and USB transceivers. The parts use a Cypress-developed USB optimized instruction set and are the first to offer EPROM programmability for fast customization. Special features include a clock doubler and Instant-On-Now® ensuring low EMI and 70% less power consumption. An additional 12-40 general purpose I/Os allow support for a variety of USB applications

The Cypress USB family consists of a set of low-speed microcontrollers, high-speed microcontrollers and hub solutions. The full USB selection is listed below:

Low Speed USB Microcontrollers (1.5Mbps)

Part Number	max. RAM	max. EPROM	# I/Os
CY7C360xx	128Bytes	4KB	12
CY7C631xx	128Bytes	4KB	16
CY7C632xx	128Bytes	4KB	10
CY7C634xx	256Bytes	8KB	32
CY7C635xx	256Bytes	8KB	40

Stand-alone USB hub solutions with I^{2}C interface (12Mbps)

Part Number	# Hub Ports	# I/Os
CY7C650xx	8	16
CY7C651xx	4	12

High Speed USB Microcontrollers with I^{2}C interface (12Mbps)

Part Number	# Hub Ports	# I/Os
CY7C660xx	4	32
CY7C661xx	4	40

Chapter 6: Examples of USB Hub and Device ICs

The Cypress USB solutions are ideal for all USB peripheral and hub applications including mice, joysticks, gamepads, keyboards, scanners, cameras, monitors, and more. The parts are manufactured in US-based wafer manufacturing plants, ensuring high quality, reliable delivery, and support.

Design Tips:

The CY365x Developer's Kit

All Cypress USB solutions are supported by the CY365x Developer's Kit. The CY365x is a full-speed hardware emulator to help you develop firmware and system drivers for Cypress USB microcontrollers. The kit contains assembly software, debug software, source code, and complete documentation. It offers user-configurable break-points, single stepping, full register, RAM, and I/O display status as well as access to key microcontroller internal signals to enable trace and complex break-points.

Example Reference Designs :

Mouse with a Cypress µC

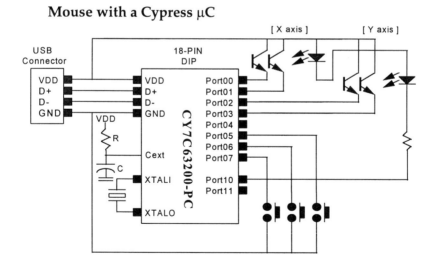

Device ICs

Joystick with a Cypress μC

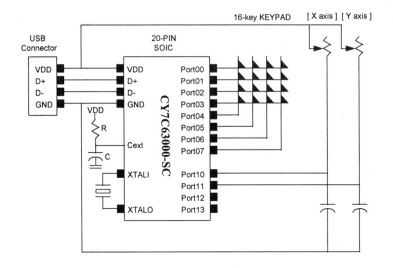

Company Name:	Cypress Semiconductor
Address:	3901 North First Street
	San Jose, CA 95134
Telephone:	1-800-858-1810
Fax:	408-943-6848
Web:	http://www.cypress.com

Company Name:	Future Technology Devices International Limited
Product Name:	USB Keyboard Controller with Integral PS/2 Mouse Support
Model Number:	FT8U48AM
Features:	

Integral low speed USB interface (UPI) with 1x8-byte deep control R/W endpoint and 2x8-byte deep R/W data endpoints.

4k byte internal mask ROM pre-programmed with FTDI's USB HID class keyboard peripheral firmware.

128 bytes internal data RAM

Optional configuration through an external 2k bit serial EEPROM (256 x 8 bit) allows different key switch matrix configurations, languages, and user / product ID codes from a standard device.

Dedicated masked ROM versions available for large volume production eliminates the need for the optional EEPROM.

18 key switch matrix drive lines.

8 key switch matrix sense lines

3 LED drivers for status indicators

EMCU microcontroller core running at 6 MHz internal CPU clock speed.

System timer provides 1ms interrupts for keyboard scan timing, communication timeouts, and error correction.

PS/2 Mouse Support (FT8U48AM-002 only) for lowest cost USB keyboard / mouse combination.

Low power USB Suspend mode.

48 pin SDIP package

Description: The FT8U48AM series of USB peripheral controllers are designed specifically for implementing cost-effective USB keyboards.

The device includes FTDI's tested firmware providing the functions of USB device enumeration, communications and error recovery, USB power management, USB HID class keyboard device protocol, key switch matrix scanning, debounce, decoding and phantom key detection. By providing standard firmware, the FT8U48AM series eliminates the need for extensive firmware development, reducing time to market and simplifying qualification of finished products to the USB standard.

There are two standard firmware variants of the FT8U48AM. The 001 version implements all the standard functions required for a USB keyboard controller design. The 002 version adds support for plugging a 2 or 3 button PS/2 mouse into the keyboard, translating the mouse movements into USB HID protocol.

There are two different methods of configuring the keyboard controller for different vendor / product IDs, key switch matrices and languages. The first uses a low-cost external EEPROM in an 8-pin DIP package. This allows a standard device to service most common keyboard designs by simply changing the contents of the EEPROM. For large volume production of specific designs, custom mask programming of the FT8U48AM is available, thus eliminating the cost of the EEPROM from the overall build cost.

For keyboard designs requiring an integral USB hub, please refer to the FT8U100AM data sheet.

Company Name:	Future Technology Devices International Limited
Address:	St George's Studios
	93/97 St George's Road
	Glasgow G3 6JA
	United Kingdom
Telephone:	(44) 141 353 2565
Fax:	(44) 141 353 2656
Email:	ftdi@msn.com
Web:	http://www.ftdi.co.uk

Company Name:	Future Technology Devices International
Product Name:	USB Mouse/Joystick Controller
Model Number:	FT8U24AM
Features:	

Integral low speed USB interface (UPI) with 1x8-byte deep control R/W endpoint and 1x8-byte deep R/W data endpoint.

4k byte internal mask ROM pre-programmed with FTDI's USB HID class mouse or joystick peripheral firmware.

Low power USB suspend mode.

2- or 3-button mouse support (version 001)

Optional Z-axis control for 3D mouse implementation (version 001)

Up to eight joystick buttons and hat switch control (version 002)

Up to four joystick analog controls supported (version 002)

EMCU microcontroller core running at 6 MHz internal CPU clock speed.

System timer provides 1 ms interrupts for peripheral timing and control, communication timeouts, and error correction.

24 pin SDIP package

Description: The FT8U24AM series of USB peripheral controllers are designed specifically for implementing cost-effective USB mice and joysticks.

The device includes FTDI's tested firmware providing the functions of USB device enumeration, communications and error recovery, USB power management, USB HID class mouse (or joystick) device protocol, key switch, and optical decoder scanning. By providing standard firmware, the FT8U24AM series eliminates the need for extensive firmware development, reducing time to market and simplifying qualification of finished products to the USB standard.

There are two standard firmware variants of the FT8U24AM. The 001 version implements all the standard functions required for a USB mouse controller design. The 002 version implements all the standard functions required for a USB joystick controller design.

The 001 mouse controller supports 2 or 3 button mice, and has the option of a 3rd optical encoder input for mice supporting a Z-Axis rotary control.

The 002 joystick controller supports up to 4 analog controls (X-Axis, Y-Axis, Throttle and Rotation), and up to 8 buttons and a hat switch through a drive/sense switch matrix.

Company Name:	Future Technology Devices International Limited
Address:	St George's Studios
	93/97 St George's Road
	Glasgow G3 6JA
	United Kingdom
Telephone:	(44) 141 353 2565
Fax:	(44) 141 353 2656
Email:	ftdi@msn.com
Web:	http://www.ftdi.co.uk

Company Name:	Lucent Technologies
Product Name:	Device Controller
Model Number:	USS-820

Features:

Full compliance with the Universal Serial Bus Specification 1.0

Supports control, interrupt, bulk, and isochronous endpoints

USB device controller with protocol control and administration for up to 16 USB endpoints

On-chip PLL allows operation at 12 Mbits/s

On-chip crystal oscillator

Integrated full-speed USB transceivers

Implemented in Lucent's 0.35 mm, 3 V standard-cell library

Applications:

Suitable for peripherals with embedded microprocessors

Glueless interface to microprocessor buses

Support of multifunction USB implementations, such as printer/scanner and integrated multimedia applications

Suitable for a broad range of device class peripherals in the USB standard

Description: USS-820 is a USB device controller that provides a programmable bridge between the USB and a local microprocessor bus. It allows any PC peripheral to upgrade to USB connectivity without major redesign effort. It is programmable through a simple read/write register interface that is compatible with industry-standard USB microcontrollers. USS-820 is designed in 100% compliance with the USB industry standard, allowing device-side USB products to be reliably installed using low-cost, off-the-shelf cables and connectors.

The integrated USB transceiver supports 12 Mbits/s full-speed operation. FIFO options support all four transfer types: control, interrupt, bulk, and isochronous, as described in USB specification 1.0, with a wide range of packet sizes. Its double sets of FIFO allow back-to-back transfer to reduce latency. FIFO sizes are programmable through control register settings. The sizes supported are 16 bytes and 64 bytes for non-isochronous pipes and 64 bytes, 256 bytes, and 512 bytes for isochronous pipes. This covers a wide range of data rates, data types, and combinations of applications. USS-820's FIFO control manager (FCM) handles the data flow between the FIFOs and the device controller. It handles flow control and error handling/fault recovery to monitor transaction status and to relay control events via interrupt vectors.

The USS-820 is available in a 44-pin MQFP package. USS-820 supports a maximum of eight bi-directional endpoints with 16 FIFOs (eight for transmit and eight for receive) associated with them. The FIFOs are on-chip, and sizes are programmable up to a total of 1 Kbyte.

The USS-820 can be clocked either by connecting a 12 MHz crystal to the XTAL1 and XTAL2 pins, or by using a 48 MHz external oscillator.

Serial Interface Engine: The SIE is the USB protocol interpreter. It serves as a communicator between the device controller and the host through the USB lines. The SIE functions include the following:

Package protocol sequencing

SOP (start of packet), EOP (end of packet), RESUME, and RESET signal detection and generation

NRZI data encoding/decoding and bit stuffing

CRC generation and checking for token and data

Serial-to-parallel and parallel-to-serial data conversion

FIFO:

The following table shows the programmable FIFO sizes.

FFSZ[1:0]	00	01	10
non-ISO	16 bytes	64 bytes	64 bytes
ISO	64 bytes	256 bytes	512 bytes

Each FIFO can be programmed independently, but the total size (TX FIFOs + RX FIFOs) must not exceed 1 Kbyte.

The transmit and receive FIFOs are accessed by the application through the register interface. The transmit FIFO is written to via the TXDAT register, and the receive FIFO is read via the RXDAT register. The particular transmit/receive FIFO is specified by the EPINDEX register. Each FIFO is accessed serially, and each RXDAT read increments the receive FIFO read pointer by 1, and each TXDAT write increments the transmit FIFO write pointer by 1.

Each FIFO consists of two data sets to provide the capability for simultaneous read/write access. Control of these pairs of data sets is managed by the hardware, and is invisible to the application, although the application must be aware of the implications. The receive FIFO read access is advanced to the next data set by setting the RXFFRC bit of RXCON. This bit clears itself after the advance is

Chapter 6: Examples of USB Hub and Device ICs

complete. The transmit FIFO write access is advanced to the next data set by writing the byte count to the TXCNTH/L registers.

The USB access to the receive and transmit FIFOs is managed by the hardware, although the control of the data sets can be overridden by the ARM and ATM bits of RXCON and TXCON, respectively. A successful USB transaction causes FIFO access to be advanced to the next data set. A failed USB transaction (e.g., for receive operations, FIFO overrun, data timeout, CRC error, bit stuff error; for transmit operations, FIFO underrun, no ACK from host) causes the FIFO read/write pointer to be reversed to the beginning of the data set to allow transmission retry for non-isochronous transfers.

The transmit FIFOs are circulating data buffers that have the following features:

Support up to two separate data sets of variable sizes

Include byte counter register for storing the number of bytes in the data sets

Protect against overwriting data in a full FIFO

Can retransmit the current data set

All transmit FIFOs use the same architecture. The transmit FIFO and its associated logic can manage up to two data sets, data set 0 (ds0) and data set 1 (ds1). Since two data sets can be used in the FIFO, back-to-back transmissions are supported.

The CPU writes to the FIFO location that is specified by the write pointer. After a write, the write pointer automatically increments by 1. The read marker points to the first byte of data written to a data set, and the read pointer points to the next FIFO location to be read by the function interface. After a read, the read pointer automatically increments 1.

When a good transmission is completed, the read marker can be advanced to the position of the read pointer to set up for reading the next data set. When a bad transmission is completed, the read pointer can be reversed to the position of the read marker to enable the function interface to reread the last data set for retransmission. The read marker advance and read pointer reversal can be achieved

109

two ways: explicitly by firmware or automatically by hardware, as indicated by bits in the transmit FIFO control register (TXCON).

The receive FIFOs are circulating data buffers that have the following features:

Support up to two separate data sets of variable sizes

Include byte count register that accesses the number of bytes in data sets

Include flags to signal a full FIFO and an empty FIFO

Can re-receive the last data set

A receive FIFO and its associated logic can manage up to two data sets, data set 0 (DS0) and data set 1 (ds1). Since two data sets can be used in the FIFO, back-to-back transmissions are supported.

The receive FIFO is symmetrical to the transmit FIFO in many ways. The SIE writes to the FIFO location specified by the write pointer. After a write, the write pointer automatically increments by 1. The write marker points to the first bye of data written to a data set, and the read pointer points to the next FIFO location to be read by the CPU. After a read, the read pointer automatically increments by 1.

When a good reception is completed, the write marker can be advanced to the position of the write pointer to set up for writing the next data set. When a bad transmission is completed, the write pointer can be reversed to the position of the write marker to enable the SIE to rewrite the last data set after receiving the data again. The write marker advance and write pointer reversal can be achieved two ways: explicitly by firmware or automatically by hardware, as specified by bits in the receive FIFO control register (RXCON).

The CPU should not read data from the receive FIFO before all bytes are received and successfully acknowledged, because the reception may be bad.

To avoid overwriting data in the receive FIFO, the SIE can monitor the FIFO full flag (RXFULL bit in RXFLG). To avoid reading a byte when the FIFO is empty, the CPU can monitor the FIFO empty flag (RXEMP bit in RXFLG).

Chapter 6: Examples of USB Hub and Device ICs

Company Name:	Lucent Technologies, Inc.
	Microelectronics Group
Address:	555 Union Boulevard
	Room 30L-15P-BA
	Allentown, PA 18103
Telephone:	800-372-2447
Fax:	610-712-4106
Email:	docmaster@micro.lucent.com
Web:	http://www.lucent.com/micro

Company Name:	Mitsubishi
Product Name:	High Speed USB 8-bit MCU
	(Also available M37532 Low Speed MCU)
Model Number:	M37640

Features:

Mitsubishi 7600 Series 8-bit CPU Core

Universal Serial Bus Specification 1.0 Compatible

On-Chip USB Transceiver

Five Transmit and Receive FIFOs

Supports High Speed Functions

Supports All USB Transfer Types: Isochronous, Bulk, Control, and Interrupt

Instruction Execution Time of 83ns (fin=24MHz)

Single Power Supply of Vcc = 5V ± 5%

ROM: 32 KB on-chip

RAM: 1K Bytes on-chip

Device ICs

DMA Controller: 2 Channels. Single Byte and Burst Transfer Modes

Timers: Three 8-bit and Two 16-bit Counters

Programmable I/O Ports: CMOS - 66 total

Dual Full Duplex UARTs. Baud Rate: 9.5 - 625Kbits/Second (at F = 10MHz)

Master CPU Bus Interface

Built-in Microprocessor and Memory-Expansion Modes

Up to 64K bytes of EPROM, SRAM or DRAM Memory can be Accessed

On-chip Count Source Generator creates a clock with user-definable frequency and duty cycle

Package: 80 pin QFP (0.8mm pin spacing)

Description: The MELPS 740 MCU architecture brings high speed, low power dissipation, efficient software support (C and/or assembly), and a wide set of integrated features to the M37640 USB function controller.

The M37640 contains a 7600 series CPU core fabricated on a 0.8 mm CMOS process. In its present form, the M376xx Series has the ability to address up to 64K bytes of memory, and possesses the following set of hardware features: 32KB of ROM and 1KB of RAM, a main clock frequency up to 24MHz, two DMA channels, 66 general purpose I/O ports, three 8-bit and two 16-bit timers, two full duplex UARTs, a frequency synthesizer, a low-power mode and the USB Function control unit with on-chip USB tranceiver.

The M37640 is Mitsubishi Electronics' first generation USB microcontroller based on the powerful 7600 series CPU core. The addition of the Universal Serial Bus (USB) core and on-chip peripherals to this highly integrated device provides the following advantages:

Ease of use (hot pluggability, compatibility of PC peripherals with each other, Plug and Play)

Chapter 6: Examples of USB Hub and Device ICs

Port expansion. Currently many PC designs have limited serial and parallel ports.

Single connector across all PC peripheral devices.

Telephony Integration Support. USB will enable many standard communication interfaces such as POTS, ISDN, PR1, T1, and E1 to exist concurrently.

The microcontroller integrates the analog transceiver eliminating the need for an external device.

The Intel 8042-compatible bus interface enables the M37640 to operate in a master/slave mode and to communicate with an external Host CPU.

An internal clock generator provides the CPU clock and the 48 MHz clock for the USB block (this also helps reduce EMI problems).

Two independent DMA channels provide an efficient means of transferring USB data between the USB FIFOs and the 8042 compatible master bus.

A low-power mode creates energy savings in Idle Mode.

Company Name: Mitsubishi Electronics America, Inc.
Address: Sunnyvale, CA, USA
Telephone: 408-730-5900.
Fax: 408-245-8214
Email: usb@msai.mea.com
Web: www.mitsubishichips.com

Company Name: Motorola
Product Name: Mouse Microcontroller
Model Number: 68HC05JB2

Description: This microcontroller is the first of a family of USB microcontrollers planned by Motorola. This chip has an on-chip memory of 2048 bytes of user ROM (or EPROM) and 128 bytes of user RAM. It also has 16-bit input capture/output compare timers and is designed to be fully compliant with low-speed USB with three endpoints. Other features include ten bi-directional I/O pins, low voltage reset (LVR) circuit, a 3.3V +/- 10% DC voltage for USB pull-up resistor, and hardware mask and flag for external interrupts. It also provides fully static operation with no minimum clock speed, multifunction timer, power-saving STOP and WAIT modes, illegal address reset, 8 x 8 unsigned multiply instruction, a 20-pin plastic dual in–line package (PDIP) and 20-pin surface mount small outline package (SOIC).

Mask options for this chip include external interrupt pins, -- edge-triggered or edge- and level-triggered; Port A and Port B pull-down and pull-up resistor – connected or disconnected; Port A PA0-PA3 external interrupt capability – enabled or disabled; OSC – crystal/ceramic resonator startup delay; crystal/resonator feedback resistor – connected or disconnected; and LVR – enabled or disabled. The chip also includes a voltage regulator.

A comprehensive line of development tools is available for this chip, including the high-performance Motorola Modular Development System (MMDS), the cost-effective Motorola Modular Evaluation System (MMEVS), as well as a programmer. A USB device firmware library supporting this chip for all low-speed USB HID applications is being planned. Prototype code is currently available for preliminary application code development with a production date targeted for 3Q97. Motorola intends to port this library to each successive device in their USB family, with functionality expanded to cover future device classes. Additionally, a line of tool support is available from third-party vendors.

Company Name: Motorola

Address: 6501 William Cannon Drive West

Austin, TX 78735

Telephone: 800-765-7795

Web: http://sps.motorola.com/csic

Chapter 6: Examples of USB Hub and Device ICs

Company Name: Nogatech, Inc.

Product Type: Video Chip

Product Name: Real-time video chip for video conferencing

Model Number: NT1003

Features:

Up to 30 frames/sec @ CIF size (352x288 pixels)

Selectable raw/compressed video out

Variable compression ratio: 2 bpp to 7 bpp

Selectable USB bandwidth (0.5 Mbps - 8.0 Mbps in 0.5 Mbps steps):

> CIF image @ 15f/s: Variable bandwidth 3.0 Mbps - 8.0 Mbps
>
> QCIF images @ 30f/s: Variable bandwidth 1.5 Mbps - 8.0 Mbps
>
> QCIF images @ 15f/s: Variable bandwidth 0.8 Mbps - 8.0 Mbps

Built-in programmable scaler

Built in zoom and pan capabilities

Supports high-resolution still image capture (640 x 480 pixels)

Full duplex audio support (using a low cost telephony codec)

Low power consumption (200mW @ 3.3V) - can use USB power source

Fits Intel's MMX concept

Description: Nogatech's exciting new development is the chip and drivers that will enable video conferencing over a USB camera. This new technology is revolutionary in its scope, as USB will be available in every future computer: desktop, notebook, PDA, and NC. The NT1003 is an ideal solution for digital camera manufacturers who want to utilize the Universal Serial Bus as an interface between the camera and the computer. Using a proprietary video compression algorithm, the NT1003 enables a throughput of up to 30 frames per second for a CIF size images, utilizing less than half of total available bandwidth of the USB port. An appropriate software driver in the host computer de-compresses the incoming USB data, consuming only 12.5% of CPU time (at CIF 15 frames per second); and the resulting digital image quality is almost identical to the source coming from the camera.

As most of the applications of a camera-computer combination require audio as well, the NT1003 also includes special logic to support cost effective full duplex digital audio. These monolithic features of the NT1003 make it ideal to be used as a one-chip solution for a very low cost, portable video conferencing application via USB.

Nogatech's chip offers the compression necessary for passing data from the camera through the USB channel into the computer. The biggest advantage of Nogatech's chip is that it is a low cost all-in-one solution, allowing for image processing, image compression, a USB interface, a CCD interface, memory interface, audio interface, and I²C interface. Possible image sizes include CIF, QCIF and Sub-QCIF at a frame rate of approximately 15 to 30 frames per second. The Video Conferencing cameras allow for conferencing over the various video conferencing standards (H.320, H.323, and H.324).

Company Name:	Nogatech Inc.
Address:	2 Mivtza Kadesh Street, Givat Shmuel, Israel 54100
Telephone:	+972-3-532-4099, ext. 128
Fax:	+972-3-532-4399
Email:	lhod@nogatech.co.il
Web:	http://www.nogatech.com

Company Name:	Samsung Semiconductor, Inc.
Product Name:	KS57C6002 Device IC
Features:	

Memory: 256 x 4-bit data memory (RAM), 2048 x 8-bit program memory (ROM)

I/O: 15 pins, including two for USB

Interrupts: one external interrupt vector, three internal interrupt vectors, two quasi-interrupts

Power Down modes: Idle mode - only CPU clock stops, stop mode - system clock stops

Timers: 8-bit programmable interval timer, 8-bit timer/counter-programmable interval timer, external event counter function, timer clock output to TIO pin

Watch Timer: Time interval generation - 0.33s, 2.16 ms at 6.0 MHz, four frequency outputs to BUZ pin

Comparator Inputs: Four-channel mode using internal reference with 4-bit resolution, three-channel mode using external reference with +/-50 mV resolution

Oscillator: Crystal/ceramic: 6 MHz (typical), CPU clock divider circuit (by 4, 8, or 64)

Eight frequency outputs to the CLO pin

Instruction execution times: 0.65, 1.33, 10.7 µs at 6 MHz (5V)

Description: The KS57C6002 single-chip 4-bit microcontroller is fabricated using an advanced CMOS process. With comparator inputs, high-current LED direct drive pins, USB I/O interface, and three versatile 8-bit timer/counters, the KS57C6002 offers an excellent design solution for USB slow speed applications such as mouse, joystick, and communications interface.

Design Tips: The SMDS (Samsung Microcontroller Development System) is a complete, PC-based development environment designed to help you concentrate on product development (not on learning debugging quirks and idiosyncrasies) so that you get the job done quickly and on time.

The SMDS is tailored for design and prototyping environments with debugging access at the source assembly language level. The SMDS contains step, break-point, and debug facilities, as well as a real-time trace of instructions executed. A trace cable connection is provided, allowing for logic analyzer type functions to be used as a debugging tool to analyze real-time MCU operations.

The SMDS comes complete with the necessary hardware, software, device-specific development files, a universal assembler, cables, and documentation.

Company Name: Samsung Semiconductor, Inc.
Address: 3655 N. First St.
San Jose, CA 95134-1708
Telephone: 408-954-7000, 800-446-2760
Fax: 408-954-7286
Web: www.sec.samsung.com

Company Name: ScanLogic Corporation
Product Name: SL11-USB Controller
Model Number: SL11
Features:

   Standard microprocessor interface

   Supports DMA transfers

   8 bit bi-directional parallel interface

   Supports multiple data types:

      32 bit color data

      True eight-bit grayscale data

      Binary data

   256 x 8 on-chip memory array

   Four USB endpoints

   On-chip USB transceiver (driver)

Driver software

Windows 95 VxD / DLL drivers, and TWAIN interface software available

5V, 0.8 micron CMOS technology

28 pin PLCC package

USB capabilities

12 Mbits/second transfer rate

Concurrent operation of multiple devices

Hierarchical topology supports 63 devices

Guaranteed service latency

Guaranteed bandwidth allocation

Supports mixed mode isochronous, asynchronous, and bulk data transfer modes

Inherent error handling/fault detection and recovery capabilities

Hot Plug and Play

Standard, Class device drivers included with operating systems

Fast page transfer rates:

> Two pages/minute @ 300 dpi (A4, 24 bit color, uncompressed)
>
> Four pages/minute @ 200 dpi (A4, 24 bit color, uncompressed)
>
> Twelve pages/minute @ 100 dpi (A4, 24 bit color, uncompressed)

Company Name:	ScanLogic Corporation
Address:	4 Preston Court
	Bedford, MA 01730
Telephone:	617-276-3901
Fax:	617-275-1758

Email:	Slinc@ix.netcom.com

Company Name:	USAR Systems
Product Name:	Function IC for Keyboards, Mice, Touch Screens and the HulaPoint
Model Number:	UR5HCUSB
	UR7HCUNI-USB
	UR7HCUSBM

Features:

  Function IC with free HID core code for Intel 930

  Provides PS/2 port for legacy keyboard and mouse devices

  UR7HCUNI offers combination Touch Screen and HulaPoint complete embedded mouse

  Works with a variety of off-the-shelf keyboards and mice

  Inexpensive single-chip mouse solution soon available

  Key matrixes may be customized to suit specific applications

  OTP and masked ROM versions available

Description: USAR Systems' USB function ICs for input devices provide a simple way to immediately access the USB. Used in a reference design by Intel, these ICs are fully compliant to the latest USB specification revision. They are low power devices, and conform to the power management requirements of the USB. In addition, the ICs provide an auxiliary port for the connection of a PS/2 compatible keyboard or mouse.

Company Name:	USAR Systems
Address:	568 Broadway, #405
	New York, NY 10012
Telephone:	212-226-2042

Chapter 6: Examples of USB Hub and Device ICs

Fax:	212-226-3215
Email:	egooch@usar.com

Company Name:	Winbond Electronics Corp.
Product Name:	W81C280 USB Keyboard Controller

Description: W81C280 is a single chip microcontroller with a USB interface for keyboard applications. It supports an 18 x 8 keyboard scan matrix, which allows suspend, wakeup, and also provides a port for a PS2 mouse. With extra endpoints it can be used for multifunction device design.

Company name:	Winbond Electronics Corp.
Address:	No. 4, Creation Rd. III
	Science-Based Industrial Park
	Hsinchu, Taiwan, R.O.C.
Telephone:	886-3-5770066, ext. 7018
Fax:	886-3-5792646
Email:	khlin@winbond.com.tw
Web:	http://www.winbond.com.tw

6.3 Host ICs

Company Name:	Cypress Semiconductor
Product Type:	Host IC
Product Name:	Peripheral Controller
Model Number:	CY82C693U

Description: The CY82C693U is peripheral controller containing a PCI-ISA bridge, an EIDE controller, a Real-Time-Clock, a Keyboard/Mouse Controller, and an OHCI USB Host Controller. It can be used as a stand-alone device or as part of the Cypress hyperCache chipset.

Company Name: Cypress Semiconductor
Address: 3901 North First Street
San Jose, CA 95134
Telephone: 1-800-858-1810
Fax: 408-943-6848
Web: http://www.cypress.com

Company Name: Intel Corporation
Product Name: 8x930 USB Peripheral Controller Family
Model Number: 8x930Ax and 8x930Hx
Product Type: Device IC (Hubless and Hub Models)
Features:

MCS® 251 Architecture Compatible Core

Complete USB 1.0 Specification compatibility

Up to six transmit/receive FIFO pairs

256 Kbyte memory addressing

Full speed (12 Mbps) and Low Speed (1.5 Mbps) subchannel USB data rate support

Suspend/Resume operation

User-selectable configurations for wait states, external address microcontroller range, and page mode

Programmable Counter Array (PCA)

Intel end-to-end USB system validation

Integrated USB Hub Repeater

Description: The USB topology has three elements (Host, Hub, and Function) that work together to enable the four different transfer types. Within a USB system, the host controls the flow of data and control information over the bus. This host capability is normally found on the PC motherboard, such as the integrated USB host found in Intel's PCIset chip sets. Functions provide capabilities to the host system. These can include typical PC activities such as keyboard or joystick input and monitor controls, or more advanced activities like digital telephony and image transfer. The Intel 8x930Ax peripheral microcontroller is designed for function control. Finally, hubs provide an expansion point for USB by supplying a connection to other USB functions. The Intel 8x930Hx, which integrates USB function and hub control features, was the first production USB hub available for today's PC peripheral devices.

USB hubs play an integral role in expanding the world of the PC user. With device connections furnished by embedded hubs in keyboards, monitors, printers, and other devices, attaching or removing a new peripheral is as simple as reaching for the plug. USB, featuring new levels of throughput and expanded connectivity sites, could even bring about many new peripherals for the next generation of entertainment and productivity applications. Long gone are the days of add-in cards, IRQ conflicts, and knotted tangles of wiring.

For even simpler connectivity, the USB cable consists of only four wires: V_{bus}, D+, D-, and GND. A single standardized upstream connector type further increases the ease-of-use of USB peripherals. The data is differentially driven over D+ and D- at a bit rate of 12 Mb/s for full-speed signaling, or a rate of 1.5 Mb/s for the USB low-speed signaling mode. The USB-compliant 8x930 family has implemented the signaling transceiver on-chip, eliminating the need for all external circuitry, except for the pull-up terminating resistor on either the D+ or D- line to determine whether the device is full- or low-speed.

Family Overview: The Intel 8x930 family includes two different single chip Universal Serial Bus 1.0 specification-compliant microcontrollers.

Intel's 8x930Ax is an 8-bit microcontroller designed specifically for USB peripherals and based on the MCS® 251 microcontroller architecture. The 8x930Hx, on the other hand, features the same time-proven MCS 251 controller core, plus the advanced capability of an integrated USB hub controller. The MCS 251 architecture of both USB microcontrollers brings with it:

High performance

Substantial memory mix and addressing

Low power

Low noise

Efficient high-level language support

Enhanced instruction set

Integrated features

Code for the 8x930Ax and 8x930Hx can use either the MCS 51 or MCS 251 microcontroller instruction sets. This gives you the option of protecting their software investment or gaining maximum performance in their application.

The 8x930 USB microcontrollers feature a rich combination of integrated peripherals that make them even more powerful. A Programmable Counter Array (PCA) provides the flexibility for applications that require real-time compare/capture, high speed I/O, and pulse-width modulation capabilities. Also included on both devices is an enhanced serial port, three 16-bit timer/counters, a hardware watchdog timer, four 8-bit I/O ports, and two power-saving modes: idle and power down.

Feature	Benefit
MCS® 251 Architecture Compatible Core	Easily migrate MCS 51/251 architecture code to preserve software investment or use MCS 251 controller code to optimize performance
Complete USB 1.0 Specification compatibility	Ensures successful and easy design of USB peripherals.
Four transmit/receive FIFO pairs	Ensures high USB packet transfer rates for standalone functions. Allows hub functions to control one internal downstream device.
256 Kbyte memory addressing	Supports increased code and data requirements

Chapter 6: Examples of USB Hub and Device ICs

Full speed (12 Mbps) and Low Speed (1.5 Mbps) subchannel USB data rate support	Easy design of USB peripherals with various data rate level requirements.
Suspend/Resume Operation	Low current required during function idle state.
User selectable configurations for wait states, external address range, and page mode	Allows the use of varied memory configurations and speeds for cost optimized designs.
8x930Hx Hub Repeater	Hub device can control up to three external downstream devices.
Programmable Counter Array (PCA)	Provides real-time capture/compare, high-speed output, and PWM functions.
Intel end-to-end USB system validation	Ensures that the 8x930 family is 100% compatible with the USB 1.0 specification and optimized to work efficiently with Intel PCIset chip sets.

Description: The 8x930 family has 1 Kbyte of on-chip data RAM and is available in ROMless, 8Kbytes ROM, and 16Kbytes ROM versions. These devices also have up to 256Kbytes of external code/data memory space and 40 bytes of general purpose registers which reside in the CPU as register files. There are 16 possible byte registers, 16 possible word registers, and 10 possible Dword registers in the register file, depending on the combinations used. Intel's USB microcontroller family provides additional flexibility when interfacing to external memory. Designers have the option of utilizing up to three additional wait states or using the real time wait function to generate more wait states, which allows the use of slower memory devices. External instruction fetches can double performance by using Page mode to swap the data onto the high byte of the address.

Both microcontrollers in the 8x930 family have eight FIFOs (first in, first out) for internal downstream device support: four transmit FIFOs and four receive FIFOs. The four transmit/receive FIFOs support four function endpoints (0-3). Endpoint 0 is 16 bytes and is dedicated for control transfers. Endpoint 1 is user configurable up to 1024 bytes, and Endpoints 2 and 3 are each 16 bytes. These three endpoints can be used for interrupt, isochronous, or bulk transfer types. In the case of the 8x930Hx, these FIFOs are augmented by a FIFO pair for upstream communications. These endpoints are supported in the 8x930Hx by an additional Repeater unit, which is responsible for re-transmitting the data streams generated by downstream devices.

Part Name	ROM Size	RAM Size	Package	Speed	Hub/ Hubless
80930AD	ROMless	1 Kbyte	68 lead PLCC	12 MHz	Hubless
83930AD	8 Kbytes	1 Kbyte	68 lead PLCC	12 MHz	Hubless
83930AE	16 Kbytes	1 Kbyte	68 lead PLCC	12 MHz	Hubless
80930HD	ROMless	1 Kbyte	68 lead PLCC	12 MHz	Hub
83930HD	8 Kbytes	1 Kbyte	68 lead PLCC	12 MHz	Hub
83930HE	16 Kbytes	1 Kbyte	68 lead PLCC	12 MHz	Hub

The implementation of USB on the 8x930Ax and 8x930Hx can be divided into four sections: FIFOs, Function Interface Unit, Serial Bus Interface Engine, and the Transceiver. The 8x930Hx has additional sections to control the additional hub functions, the Hub Interface Unit, and the Repeater.

The transmit and receive FIFOs on both devices are circulating FIFOs supporting up to two separate data sets of variable sizes and containing byte count registers to access the number of bytes in the data sets. They also have flags that detect a full or empty FIFO and have the capability of re-transmitting or re-receiving the current data set.

The Function Interface Unit (FIU) manages the USB data received and transmitted based on the transfer type and the state of the FIFOs. It is responsible for monitoring the transaction status, managing the FIFOs, and relaying control events to the 8x930 CPU via interrupt requests.

The Serial Bus Interface Engine (SIE) handles the communication protocol of USB by packet sequencing, signal generation/detection, CRC generation/checking, NRZI data encoding/decoding, bit stuffing, and packet ID (PID) generation/decoding.

Chapter 6: Examples of USB Hub and Device ICs

The integrated transceiver on Intel's USB microcontrollers conforms to the simple four-wire interface defined by the USB 1.0 specification. The 8x930 family of controllers also has three interrupts associated with USB. These occur for any start of frame, transmit/receive done for function endpoints, and global suspend/resume.

On the 8x930Hx hub device, the Hub Interface Unit (HIU) serves to control and manage the status of and communication to and from the downstream ports. The hub Repeater, on the other hand, manages the propagation of signals in both directions for both the upstream and downstream USB ports.

Development Tools: Intel's USB product line is complemented by a complete set of hardware and software tools from Intel and leading third-party tools vendors. Software development is supported by ANSI C cross compiler, assembler, linker/locater, debugger and simulators from Keil Software, Production Languages Corporation, and Tasking BV. Device driver and peripheral firmware development is supported by American Megatrends, Phoenix Technologies, and SystemSoft Corporation. Hardware debug and design is supported by In-Circuit Emulators and ROM Emulators from Nohau Corporation, Metalink Corporation, and iSYSTEMS. The debugging of USB communications is supported by the USB Protocol Analyzer tool from Computer Access Technology Corporation.

Evaluation of the 8x930 USB peripheral controller family is supported by the 8x930 USB Evaluation Kit from Intel. This kit consists of a hardware evaluation board that supports both the 8x930Ax and 8x930Hx devices. The board has both upstream and downstream ports, comes with documentation on the microcontroller and the board, Intel's ApBUILDER software, and evaluation C cross-compiler tools from third-party vendors.

Features:

The following features are standard on both the 8x930Ax and 8x930Hx USB peripheral microcontrollers:

Complete Universal Serial Bus 1.0 specification compatibility

- Supports isochronous and non-isochronous data

- Bi-directional half-duplex link

On-chip USB transceiver

Serial Bus Interface Engine (SIE)

- Packet decoding/generation
- CRC generation and checking
- NRZI encoding/decoding and bit stuffing

Four transmit FIFOs

- Three 16-byte FIFOs
- One configurable FIFO (up to 1Kbyte)

Four receive FIFOs

- Three 16-byte FIFOs
- One configurable FIFO (up to 1Kbyte)

Automatic transmit/receive FIFO management

Suspend/Resume operation

Three USB interrupt vectors

- USB Function Interrupt
- Start of Frame/ Hub Interrupt (8x930Hx only)
- Global Suspend/Resume

Phase lock loop

- 12 Mbps or 1.5 Mbps data rate

Low clock mode

256-Kbyte External Code/Data memory space

Power-saving Idle and Power-down modes

User-selectable configurations

- External wait state
- Address range
- Page mode

Chapter 6: Examples of USB Hub and Device ICs

Real time wait function

1 Kbyte on-chip data RAM

On-chip ROM options

- ROMless, 8 Kbyte or 16 Kbyte

Four input/output ports

- One open drain port

- Three quasi bi-directional ports

Programmable counter array (PCA)

- Five capture/compare modules

Industry standard MCS 51 microcontroller UART

Hardware watchdog timer

Three flexible 16-bit timer/counters

Code compatibility with MCS 51 and MCS 251 microcontrollers

Register-based MCS 251 microcontroller architecture

- 40-byte register file

- Registers accessible as bytes, words, or double words

6 or 12 MHz operation

In addition, the 8x930Hx contains the following hub-specific features:

USB hub

- One upstream, one internal downstream, and three external downstream ports

- Serves as both USB Hub and USB embedded function (through internal downstream port)

USB hub management features

- Connectivity management

- Downstream device connect/disconnect detection

Hub ICs

      - Power management, including Suspend/Resume

      - Bus Fault Detection and Recovery

      - Full- and low-speed downstream device support

Output pin for port power switching

Input pin for overcurrent detection

Company Name:	Intel Corporation
Address:	CH6-231
	5000 W Chandler Blvd.
	Chandler, AZ 85226-3699
Telephone:	(602) 554-1937
Web:	http://developer.intel.com/design/usb

6.4 Hub ICs

Company Name:	Alcor Micro, Inc.
Product Name:	USB Hub Controller
Model Number:	AU9204

Features:

    Single chip USB Hub with one upstream, and 4/6 downstream ports

    Compliant with USB Specification 1.0 and USB Hub Specification 1.09

    Connectivity management

    Full- and low-speed downstream device support

    Downstream device connect/disconnect detection

    Power management including suspend and resume

    Fault detection and recovery

Chapter 6: Examples of USB Hub and Device ICs

Integrated USB transceiver

Serial bus interface engine (SIE) for packet decoding/generation

CRC generation and checking

NRZI encoding/decoding and bit stuffing

Build in pair of transmit and receive FIFOs

Input pin for overcurrent sensing

Output in for port power switching

Microcontroller based architecture for easy customization to derivative products

Integrated hub and device firmware for USB compound device (e.g. keyboard and USB hub)

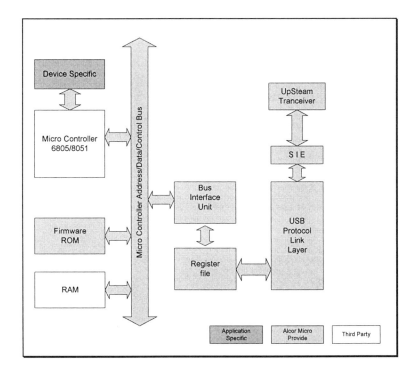

Company Name: Alcor Micro Inc.

Address:	155A Moffet Park Dr.
	Suite 240
	Sunnyvale, CA 94089
Telephone:	408-541-9700
Fax:	408-541-0378
Email:	Dennis@Alcormicro.com

Company Name:	Atmel Corporation
Product Type:	Hub IC
Product Name:	USB Hub Chip With Four Downstream Ports
Model Number:	AT43311

Features:

Self-powered hub with bus-powered controller

Full compliance with USB Spec. Revision 1.0

Downstream support for full- and low-speed transfer rates

On-chip low- and full-speed USB transceivers

Four downstream ports

6 MHz external clock for reduced EMI

Individual port control

On-chip overcurrent detection

USB connection status indicators

Description: The AT43311 is a 32-pin hub chip that can be used to create a low cost hub with optimal performance and control.

The AT43311 provides a low cost solution by integrating standard analog and digital features on-chip. In addition to reducing the solution cost, this integration saves board real estate by reducing the part count. Consider the reference voltage needed to indicate overcurrent problems. By using an on-chip analog comparator, USB designers can define the reference voltage with only a simple resistor divider.

Performance and control are not compromised by this low cost approach. The hub repeater and hub controller manage the bus activity without the need for external intervention. The hub controller also manages the status and power of each downstream port. By allowing the hub controller to power down and wake up each port individually, the system can reduce the power needs. System reliability also improves with individual port control. In the event of a power short on a peripheral, only the affected port will be shut off automatically. The remaining ports continue operation unaffected by the short.

To make the peripheral devices easier to use for the end user, the AT43311 uses port status indicators. These indicators show the customer whether a port is properly attached. A port may not be active if a short has shut off the device was not properly plugged-in.

Design Tips:

1. Noise will always be an issue for commercial products. The AT43311 reduces noise by using a low frequency 6 MHz external crystal. Internally, the clock is multiplied to 48 MHz for speed-critical areas.

To further reduce noise, standard noise reduction techniques may be followed, e.g., ferrite beads, reduction of ground loops, and proper decoupling.

2. To help speed development, designers can use the AT433DK11. This design kit comes with a stand-alone hub and an application note that details the reference design.

Company Name:	Atmel Corporation
Address:	2325 Orchard Parkway
	San Jose, CA 95131
Telephone:	408-487-2605

Hub ICs

Fax:	408-487-2600
Email:	jtatsukawa@atmel.com
Web:	www.atmel.com

Company Name:	Cypress Semiconductor
Product Name:	Hub IC
Model Number:	CY7C65xxx

Description: The CY7C65xxx is a family of hub devices for 4-8 port applications. The devices also contain an I²C interface and 12-16 general purpose I/Os.

Company Name:	Cypress Semiconductor
Address:	3901 North First Street
	San Jose, CA 95134
Telephone:	1-800-858-1810
Fax:	408-943-6848
Web:	http://www.cypress.com

Company Name:	Future Technology Devices International
Product Name:	USB Hub and Compound Peripheral Device Controller
Model Number:	FT8U100AX

Features:

Integral high speed USB interface (UPI) with 3x8-byte deep control R/W endpoints and 3x8-byte deep R/W data endpoints providing control and data communication for the hub and up to two compound functions.

Up to 7 USB downstream ports with individual power control, overcurrent detect and status indicators on each port.

Up to 35 individually programmable I/O ports available for peripheral applications.

Legacy device support - two ports for connecting PS/2 keyboards and mice.

Serial I/O port with baud rate generator and IrDA compatible mode for remote links at up to 115k baud.

Internal PLL clock multiplier provides high speed 48mhz clock from a 12MHz crystal.

EMCU microcontroller core running at 12MHz internal CPU clock speed.

External ROM (up to 32k bytes)

128 bytes internal data RAM

System timer provides 1ms interrupts for timing, control, and error correction.

Low power USB Suspend mode.

100 pin PQFP package

Firmware is available from FTDI for a wide variety of hub and peripheral devices.

Description: The FT8U100 AX USB Hub / Peripheral controller is designed to implement a wide range of USB hub controllers and compound devices, which consist of one or more functions combined with a USB hub.

Because the firmware is in an external ROM, the device can support a wide variety of USB hub configurations with up to seven downstream ports, bus- or self-powered, gang switched or individually switched power control and overcurrent indication. A wide variety of hub status reporting is accommodated with enough I/O ports being available to control two status LEDs on each of the seven possible downstream ports.

Hub ICs

Included within the FT8U100AX are two PS/2 ports for legacy PS/2 keyboard & mouse support, a UART with baud rate generator for serial I/O or IrDA communications to 115k baud.

Application areas include a wide variety of USB hubs – bus-powered or self-powered USB hubs, USB keyboard/hubs, monitor/hubs, or stand-alone hubs. Optional device support includes PS/2 keyboard, PS/2 mouse ports, and IrDA CIR port for remote infrared game controllers, mice, remote controls etc.

The FT8U100AX is recommended for use in low / medium volume applications and applications where the ability to upgrade the firmware to cope with specification changes is essential. For high volume applications where the function is well-defined and tested, FTDI will be making available the FT8U64AM device which has the same functions as the FT8U100AX but with the firmware programmed into the internal 8k mask ROM of the device.

Company Name:	Future Technology Devices International Limited
Address:	St George's Studios
	93/97 St George's Road
	Glasgow G3 6JA
	United Kingdom
Telephone:	(44) 141 353 2565
Fax:	(44) 141 353 2656
Email:	ftdi@msn.com
Web:	http://www.ftdi.co.uk

Company Name:	KC Technology, Inc.
Product Name:	USB Monitor Hub Controller
Model Number:	KC82C168
Features:	

Chapter 6: Examples of USB Hub and Device ICs

PC monitor-hub supports four down-stream USB ports and controls monitor functions

Proven compliance with USB Specification Rev. 1.0 and USB Monitor Control Device

Fully integrated microprocessor

Cost-effective solution, complete with hardware, firmware, monitor drivers, and technical support

Includes an easy-to-use screenSet applet to help users configure and control a USB monitor

Available in 64 SDIP packages

Benefits:

Offers PC users uncluttered USB connectivity that is convenient and easy to access along with Plug and Play monitor controls

Eliminates compliance headaches and helps designers get their USB-ready monitors to market sooner

Lowers development cost and time by eliminating the need to either add a dedicated microcontroller or adapt an existing microcontroller

Assures a high quality USB compliant design implementation, reducing development overhead for low cost entry to market

Lets multiple users customize, save, and restore their individual monitor settings with no manual adjustment

Space-efficient packaging offers design flexibility

Description: A Universal Serial Bus (USB) hub simplifies and extends the peripheral connectivity to the PC while hiding its complexity from the user. Easy USB plug access makes the PC monitor a perfect place for the USB hub. Thanks to USB, the next generation of PC monitors promises to be Plug and Play peripherals, delivering a new dimension in functionality, convenience, and user-friendliness. Kawatsu Corporation's KALEIDA KC82C168 USB Monitor-Hub Controller enables you to design, build, and deliver leading edge USB-capable monitors in the shortest possible time.

Hub ICs

PC monitors designed with KC82C168 provide four downstream ports to effortlessly plug digital cameras or other USB peripherals into the computer. Fully compatible with USB Specification Rev. 1.0, the KALEIDA KC82C168 USB hub with embedded monitor control is a compound four-port USB hub device. Acting as a USB hub controller, the KC82C168 device records the status of the hub, bus enumeration, and the downstream ports. The monitor controller uses an I²C style bus to enable a two-way communication between the hub and the monitor system controller.

The KC82C168 features screenSet, an easy-to-use applet running on the PC host, to help users adjust a broad range of monitor settings to suit their personal tastes and match their application needs. Kawatsu's screenSet applet makes cumbersome front-panel monitor controls a thing of the past by allowing multiple users to create, save, and reload their custom monitor settings.

The KC82C168 is a total solution for a USB monitor design, with hardware, firmware, device drivers, monitor control application, and technical support. As your USB partner, Kawatsu is committed to helping you win in the marketplace with a fast, cost-effective, and highly differentiated USB monitor-hub solution.

Company Name: KC Technology

Address: 1900 McCarthy Blvd., Suite 300

Milpitas, CA 95035

Telephone: (408) 232-9828

Fax: (408) 232-9829

Company Name: KC Technology

Product Name: USB Hub Controller

Model Number: KC82C160

Features:

State-of-the-art hub controller for Universal Serial Bus (USB) devices

Flexible support for the bus-powered, self-powered, and hybrid-powered modes

Cost effective solution in 0.5 micron CMOS technology, packaged in a 32 DIP and 44 PQFP.

One upstream and four downstream ports with dedicated power, on/off control, and overcurrent detection or ganged mode power configuration

Downstream ports support full speed and low speed peripherals

Full technical support including samples, databook, evaluation boards, and schematics

Benefits:

Highly integrated USB hub controller compliant with USB Specification Revision 1.0

Preferred power configuration selectable by designer for hub application

Enables implementation of hubs at low cost for fast gains of market share and high volume production support

Controller supports most optimal port configurations with selectable port power options matching application needs

Maximizes flexibility to support any speed peripheral offered

Fully compliant hub implementation in the shortest possible time

Description: A hub consists of two components: a hub repeater which is responsible for connectivity setup and tear-down, and a hub controller which provides the mechanism for host-to-hub communications.

The KC82C160 is fully compliant with USB Specification Rev. 1.0 and supports dynamic (or hot) attachment and removal, allowing peripherals to be attached, configured, used, and detached while the host and other peripherals are in operation. KC82C160 provides four downstream ports and one upstream port. It allows the hub to comprehend the following downstream port states on a per-port basis: connect/disconnect detect, port enable/disable, suspend/resume, reset, and power switching.

For reasons of safety, KC82C160 implements overcurrent detection over all downstream ports on a per-port basis. Hub fault recovery operates only in the upstream direction. KC82C160 detects connectivity faults, especially those that might result in a deadlock, and prevents them from occurring again.

The device is made up of the following key blocks: USB I/O Transceiver/Buffer, Serial Interface Engine (SIE), Hub Endpoint Management, and Hub Repeater.

USB I/O Transceiver/Buffer is fully compliant with electrical characteristics stipulated in the USB Specification Rev. 1.0. The hub controller can differentiate between full-speed and low-speed USB devices either when a device is connected to the bus or at power-up. It provides transceivers/buffers for both full-speed devices (12 Mbps) and low-speed devices (1.5 Mbps) The USB upstream hub port always uses full-speed signaling and its transceiver/buffer always operates with full-speed edge rates. For downstream hub ports, the speed and edge rates are selectable, depending on whether the detected device is a full- or low-speed device.

The Serial Interface Engine (SIE), which is an interface to the hub endpoint management and repeater blocks, handles the serialization and de-serialization of USB transmissions. It performs packet framing and protocol sequencing, NRZI codec, bit stuffing, CRC generation/check, packet ID generation/decode, serial-to-parallel/parallel-to-serial conversion, and sync-pattern generation.

The Hub Endpoint Management performs the system, buffer, and power management functions for the USB hub. A USB hub has two endpoints: Endpoint 0 - Control Endpoint (supports bi-directional flow of packets) and Endpoint 1 - Port Status-Change Endpoint (supports interrupt transfers).

The Hub Repeater is a protocol-controlled switch between the upstream port and the downstream ports. It has hardware support for reset and suspend/resume signaling. It also provides full-speed/low-speed recognition, detection of device attach/detach, and Loss of Activity (LOA) and babble detection.

Company Name: KC Technology

Address: 1900 McCarthy Blvd., Suite 300

Milpitas, CA 95035

Telephone: (408) 232-9828

Fax: (408) 232-9829

Company Name: Lucent Technologies, Inc.

Product Name: USS-620 USB Data Transport DMA Bridge

Features:

Full compliance with the Universal Serial Bus Specification 1.0

Supports control, interrupt, bulk, and isochronous endpoints

USB device controller with protocol control and administration for eight USB endpoints

USB data supported via data-transfer channels directly to 8-bit or 16-bit system memory

Five transmit and four receive data-transfer channels

On-chip PLL allows operation at 12 Mbits/s

Integrated full-speed USB transceiver

Microprocessor bus interrupt controller with separate device status and data-transfer reports

Programmable address mapping of data structures via descriptors and data buffers

Bus arbitration logic with low interrupt latency support

Microprocessor bus master capability for efficient data transfer to/from system memory

Hub ICs

5 V tolerant I/O buffers allows operation in 3 V or 5 V system environment.

Applications:

DMA Bus Arbiter

This module monitors internal bus requests (USB traffic) from the system control module. If a request is active, the arbiter module requests the microprocessor bus. The bus arbiter will remain bus master for a programmable number of bus transactions. If an internal request is still active after bus mastership is released, the bus master will re-arbitrate for bus mastership after a programmable number of processor clock cycles. The bus arbiter will maintain bus ownership or re-arbitrate for bus mastership until bus clear is received, no internal requests are active or the transaction counter has decremented to zero.

Interrupt Controller

This module monitors event and error interrupt requests received from the system control module. If an interrupt request is present, an IRQn signal is issued on the microprocessor bus. During the interrupt acknowledge cycle, the interrupt control module will issue the interrupt vector.

Bus Slave Decoder

This module monitors cycles on the microprocessor bus waiting to qualify an internal register address for slave read/write transactions. When an address is qualified, the registers address and strobes are sent to the channel control and status modules.

Channel Administration

The channel administration module manages nine data transporter channels to support four types of transfers in the USB architecture: control, interrupt, bulk, and isochronous. Since USB is a host-driven serial bus, it means that all bus activity is controlled and directed by the host to support information exchange between host software and a particular endpoint on a USB device. A given USB device could have an endpoint which would support a pipe for transporting data to the device and another endpoint which would support a pipe for transporting data from the USB device. See the

USS-620 System Operation overview section for the association between an endpoint and a data transport channel.

Data transmitted to/from the USB associated with each transport channel is stored in buffers located in system memory. Each channel is referenced by three elements in the channel administration module of the USS-620: channel control register, channel status register, and a channel buffer. The data stored in these registers by the processor is essential to inform the USS-620 the location, size, and type of USB data stored in external system memory and whether an endpoint is ready to receive (or transmit) this USB traffic data.

Channel Buffers

This module contains processor read/write registers, which hold the buffer information for each channel. Channel buffers contain information that describes the location and size of the USB data in system memory.

Channel Control

This module holds the registers that control the channel. The channel control is used to enable a USB endpoint and control USB-specific responses such as ACK, NAK, and STALL.

Channel Status

This module holds the registers providing status information about the channel (e.g., status of channel completion, USB packet conclusion, channel error conditions, etc.).

Bus Address/Data Multiplexer

The bus interface module provides a connection between various modules and the microprocessor bus. It provides the multiplexing functions for data flows to and from the internal FIFO storage elements of the data transporter block.

System Control

This central module maintains a state machine to control information flow through the device.

Bus Master Engine

This module is active when the USS-620 becomes the bus master and is transferring data between the system memory and the FIFOs. The address and transfer count is loaded from the channel module and the FIFOs are read from or written to using data from the bus multiplexer module.

FIFO

The FIFO module provides an elastic storage mechanism for storing USB traffic data on its way to and from system memory. It also provides flags and handshake signals to monitor system activity and control data transfers.

USB Device Control (UDC) Mux

This is the multiplexer between the FIFOs and the UDC module which terminates the USB protocol. It is responsible for routing data between the application bus of the UDC module and the correct FIFO in the data transporter.

UDC Decoder

This module decodes addresses on the UDC application bus for selecting the correct FIFO.

Signal Information

Functional Group	Signals	Number of Signals
Clocks	M68KCLK (16 MHz or 25 MHz)	1
System Control	RstN	1
Address Bus	Add[23:1], Add[0]/UDsN	24
Data Bus	Data[15:0]	16
Bus Control	Fcode[2:0], LDsN/DsN, R_WN, AsN, CsN, ENABLEN, DtackN, MemW	10
Bus Arbitration	BerrN, BrN, BGackN, BgN, BClrN	5
Interrupt Control	IrqN_ev, IrqN_er	2
Rate Match	ExtClk	1
USB	DMNS, DPLS, EXTXVER, RCV_TST_TXEN, VM_TST_LS, VP_TST_A, TXENL_TST_RCV, TXDMNS_TST_VM, TXDPLS_TST_VP, PULLUP, RstOutN, SOF	12
Test	SCAN_MOD	1
USB Clock	BYPASS, XHI, XLO	3
Power/Ground	—	24
Total		100

Chapter 6: Examples of USB Hub and Device ICs

Company Name:	Microelectronics Group
	Lucent Technologies, Inc.
Address:	555 Union Boulevard
	Room 30L-15P-BA
	Allentown, PA 18103
Telephone:	800-372-2447
Fax:	610-712-4106
Email:	docmaster@micro.lucent.com
Web:	http://www.lucent.com/micro

Company Name:	MultiVideo Labs
Product Type:	Hub/Monitor IC
Product Name:	Monitor Communication IC
Model Number:	W82C610

Description: The W82C620 acts as a USB function, a Hub controller, and a Hub repeater. It directs all traffic through the Hub between the host, four downstream ports, and the monitor. The W82C620 contains two function endpoints and two Hub endpoints for Control and Interrupt Transfers between the host and monitor/hub.

Company Name:	MultiVideo Labs, Inc.
Address:	29 Airpark Road
	Princeton, NJ 08540
Telephone:	609-497-1930
Fax:	609-497-1945
Email:	www@mvl.com
Web:	http://www.mvl.com

Company Name: National Semiconductor
Product Name: USB Hub and Monitor Controller
Model Number: LM1050

Features:

USB Specification Rev. 1.0 compliant

Supports USB display device class

USB downstream ports

– Four standard FS/LS USB ports

– Additional ports may be emulated in firmware

Access to USB hub control registers

CRC generation and checking

Five integrated USB transceivers

– 1 upstream

– 4 downstream

Integrated 3.3V transceiver regulator

I²C-compatible or UART interface

Supports up to 400kHz I²C-compatible data rates.

Programmable UART baud rate: default 9600 baud, maximum 750kbps.

Single crystal for USB hub and microcontroller.

Programmable clock output including the following frequencies: 48, 24, 16, 12, 8, 6, and 4MHz

Power enabling and current sensing interface

Supports low-cost single sided PCB-design

–Additional V_{cc} and Ground pins

–Pin-out facilitates single sided layout

28-pin dual-in-line package

Description: The LM1050 contains a microcontroller and suitable firmware to provide a full-featured USB compliant hub and display class function. The LM1050 is a mixed signal CMOS device that is compliant with USB Specification Rev. 1.0.

The USB hub has four USB downstream ports. Additional ports may be emulated in firmware. The USB upstream and downstream ports have integrated transceivers and an integrated 3.3V transceiver voltage regulator.

Either an I²C compatible or UART interface is used to support hub and monitor control. The microcontroller sends/receives data from the hub registers to perform USB hub functions. Data related to monitor control is also transferred and used by the microcontroller for monitor functions. The microcontroller interface is selected with the mode-select pin. The two interface options allow the LM1050 to be used in different microcontroller -based applications.

The UART baud rate is programmable with a default of 9600 and a maximum of 750 kbaud. A programmable clock generator provides a clock output for the microcontroller. This clock can be used to replace the microcontroller crystal. A power enable and current sense interface is provided for aggregate port power control. Support for low cost single-sided PCB design is supplied through both additional power and ground pins and an optimized pin-out.

Hub ICs

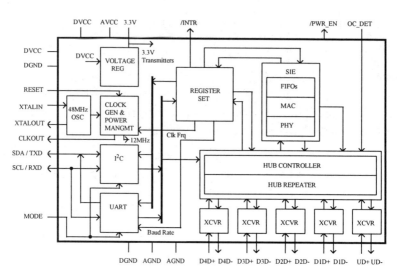

Company Name:	National Semiconductor
Address:	2900 Semiconductor Drive
	Santa Clara, CA 95052-8090
Telephone:	408-721-8762
Fax:	408-721-5691
Email:	harry.inia@nsc.com

Company Name:	NetChip Technology, Inc.
Product Name:	Four-Port USB Hub Chip
Model Number:	Net2868

Features:

Self-powered or bus-powered USB hub with a bus-powered controller. Supports both full-speed and low-speed devices

One upstream port and four downstream ports

Integrated hub protocol controller for power control, port connect / disconnect, suspend, and fault recovery

Ganged or individual port power control support

Global or individual current protection support

Enable and activity status for each port

Serial Bus Interface Engine (SIE) for packet decoding/generation, CRC generating and checking, NRZI encoding/decoding, and bit stuffing

Complete Universal Serial Bus 1.0 Specification compatibility

48-pin PQFP

Description: The NET2868 Four-Port USB Hub Chip allows up to four USB devices or hubs to connect to an upstream USB Hub or Host port. The two main components of the hub are the Repeater and the Controller.

The Repeater is responsible for the following functions:

- Connectivity setup and tear-down.
- Device connect/disconnect detection
- Suspend/resume support
- Frame synchronization
- Full/low speed device support
- Bus fault detection and recovery
- Power management

The Controller is responsible for the following functions:

- Host to hub communication
- Control registers
- Status ports
- Standard and class specific control endpoint
- Enumeration support

Company Name: NetChip Technology, Inc.
Product Name: USB Interface Controller
Model Number: NET2888

Features:

USB Specification Version 1.0 compliant

Bridges between a processor-independent local bus and a USB bus

USB device bandwidth of up to 12 Mb/sec

USB bulk, isochronous, interrupt, and control transfers.

Independent 64 byte transmit and receive FIFOs to maximize throughput.

Supports local CPU or DMA data transfers.

Low power CMOS in 48-pin plastic QFP package

3.3 V operating voltage

Description: The NET2888 USB Interface Controller allows bulk or isochronous data transfers between a generic local bus and a Universal Serial Bus (USB). The NET2888 supports the connection between a host computer and an intelligent peripheral such as a digital camera or scanner.

The three main components of the NET2888 are the USB Bus Interface, the dual 64 byte FIFOs, and a Local Bus Interface.

The USB Interface is responsible for the following functions:

    Host to device communication

    Bulk or isochronous endpoints to access FIFOs

    Interrupt endpoint to access local to USB mailboxes

    Bulk endpoint to access USB to local mailboxes

The Local Bus Interface is responsible for the following functions:

    FIFO control

    Local CPU interface

Chapter 6: Examples of USB Hub and Device ICs

    Local DMA controller interface

    Interrupts

Company Name:	NetChip Technology, Inc.
Address:	635-C Clyde Avenue
	Mountain View, CA 94043
Telephone:	415-526-1490
Fax:	415-526-1494
Email:	dty@worldnet.att.net
Web:	http://www.netchip.com

Company Name:	Philips Semiconductor
Product Name:	Hub and Monitor Controller
Model Number:	PDIUSBH11

Features:

  Complies with Universal Serial Bus Specification Rev.1.0

  Four downstream ports with per-packet connectivity

  Embedded function with two endpoints (control and interrupt)

  Integrated FIFO memory for hub and embedded function

  Automatic protocol handling

  Versatile I^{2}C interface

  Allows software control of monitor

  Compliant with USB human interface and display device class

  Single 3.3V supply

  Available in 32-pin, SDIP package

Description: The Philips PDIUSBH11 is a compound USB hub IC (hub plus embedded function). It is used in a microcontroller-based system and communicates with the system microcontroller over the I²C serial bus. This modular approach to implementing a hub and embedded function allows the designer to maintain the system microcontroller of choice and retain existing architecture. This cuts down development time and offers the most cost-effective solution. Ideal applications for the IC include computer monitors and keyboards.

The PDIUSBH11 conforms to the USB specification 1.0 and the I²C serial interface specification. It is also compliant to the USB Human Input Device and Monitor Control Class specifications.

The internal blocks of the PDIUSBH11 include the analog transceivers, the hub repeaters, frame timers, port controllers, Philips USB Serial Interface Engine, Memory Management Unit, and I²C interface.

The analog transceivers pass data through the hub repeater, which manages connectivity on a per-packet basis. Both full-speed (12 Mbit/s) and low-speed (1.5 Mbit/s) data rates are supported. The hub repeater implements packet signaling connectivity and resume connectivity. Error conditions in the hub repeater (i.e., babble error) are detected by the End of Frame timers. The timers also maintain the low-speed keep-alive strobe, which is sent at the beginning of a frame.

Via the I²C interface, a microcontroller can access the downstream ports and request or change the status of each individual port by communicating with its individual port controller. Any changes in the status or setting of the individual port will result in an interrupt request.

Data is extracted by the Philips SIE, which implements the full USB protocol layer. It is completely hardwired for speed, and needs no firmware intervention. The functions of this block include synchronization pattern recognition, parallel/serial conversion, bit stuffing and destuffing, CRC checking and generation, PID verification/generation, address recognition, and handshake evaluation/generation.

The Memory Management Unit (MMU) and integrated RAM compensate for the large difference in data-rate between USB of

12Mbps bursts and the I²C interface to the microcontroller running at 100Kbps. The I²C interface is a slave implementation allowing simpler micro-coding. The I²C interface can also be easily emulated in software for microcontrollers without I²C capability.

Company Name:	Philips Semiconductors
Address:	811 E. Arques Ave.
	Sunnyvale, CA 95066
Telephone:	408-991-3276
Fax:	408-991-2133
Email:	teresa.hardy@sv.sc.philips.com
Web:	www.semiconductors.philips.com

Company Name:	Texas Instruments
Product Name:	Four-Port Hub with Embedded Function
Model Number:	TUSB2140

Features:

Universal Serial Bus (USB) Version 1.0-compatible

Includes Serial Interface Engine (SIE)

All four downstream ports support full-speed and low-speed operations

Embedded function is USB Display Device Class compatible

Embedded function includes an I²C local bus interface

Integrated USB transceivers

Power switching and overcurrent reporting provided per-port or ganged

Available in 40-pin DIP package and 64-pin LQFP package

3.3 V operation

Hub ICs

40 MHz crystal or oscillator input

Description: The TUSB2140 hub functionality supports power switching to the downstream ports whether individually or ganged. An external device or devices are required to switch power and to detect overcurrent conditions. The TUSB2140 provides outputs to control power switching and inputs to monitor any overcurrent conditions. In the ganged mode, all PWRON signals switch simultaneously.

A crystal supplies the 48 MHz (four times full-speed USB bit rate) clock through the XTAL1 and XTAL2 pins. The SIE uses this clock to sample data from the upstream port and generate a synchronized 12 MHz USB clock signal.

USB-compliant transceivers are provided for the upstream port and all downstream ports. Every downstream port supports both full- and low-speed connections by automatically setting the slew rate according to the speed of the device attached to the port.

Company Name: Texas Instruments
Address: Post Office Box 655303
Dallas, TX 75265
Telephone: 972-644-5580
Fax: 972-480-7800
Web: http://www.ti.com

Company Name: Texas Instruments
Product Name: TUSB2040 4-Port Hub for USB
Model Number: TUSB2040
Features:

Universal Serial Bus (USB) Version 1.0 Compatible

Includes a Serial Interface Engine (SIE)

Integrated UBS transceivers

Chapter 6: Examples of USB Hub and Device ICs

Four downstream ports

Two power source modes

Self-powered mode

Bus-powered mode

Power switching and overcurrent reporting is provided per-port or ganged

All downstream ports support full-speed and low-speed operations

Supports suspend and resume operations

Available in 28-Pin DIP package and a 48-pin TQFPT package

3.3-V operation

Description: The TUS2040 hub is a CMOS device that provides up to four downstream ports in conformance with the USB specification, Version 1.0. It supports two power source modes: bus-powered and self-powered. The hub and downstream ports share the power source. The TUSB2040 hub powers down to 500 µA during the suspend operation by powering down the internal oscillator.

The TUSB2040 hub supports power switching to the downstream ports either individually or ganged. An external device or devices are required to switch power and to detect overcurrent conditions. The TUSB2040 provides outputs to control power switching and inputs to monitor any overcurrent conditions. In the ganged operation, all PWRON signals transition simultaneously, and the OVRCUR inputs should be tied together and driven by the same signal.

A crystal supplies the 48-MHz (four times the full-speed USB bit rate) clock through XTAL1 and XTAL2. The hub uses this clock to sample data from the upstream port and to generate a synchronized 12-MHz USB clock signal. USB compatible transceivers are provided for the upstream port and all downstream ports. Every downstream port supports both full-speed and low-speed connection by automatically setting the slew rate according to the speed of the device attached to the port.

Product Name: TUSB2070 7-Port Hub for the USB

Model Number: TUSB2070

Features:

Universal Serial Bus (USB) Version 1.0 Compatible

Includes Serial Interface Engine (SIE)

Integrated USB transceivers

Two power source modes

Self-powered mode

Bus-powered mode

Power switching and overcurrent reporting is provided per-port or ganged

All downstream ports support full-speed and low-speed operations

Supports suspend and resume operations

Available in 48-pin TQFPT package

3.3-V operation

Description: The TUSB2070 hub is a CMOS device that provides up to seven downstream ports in conformance with the USB specification, version 1.0. It supports two power source modes: bus-powered and self-powered. The hub and downstream ports share the same power source. When operating on bus-power, only four downstream ports are utilized. Self-powered mode is required when utilizing more than four ports as defined by the USB specification. The TUSB2070 hub powers down to 500uA during the suspend operation by powering down the internal oscillator.

The TUSB2070 hub supports power switching to the downstream ports either individually or ganged. An external device or devices are required to switch power and to detect overcurrent conditions. In the ganged operation, all PWRON signals transition simultaneously and the OVRCUR inputs should be tied together and driven by the same signal.

A crystal supplies the 48-MHz (four times full speed USB bit rate) clock through XTAL1 and XTAL2. The HUB uses this clock to sample data from the upstream port and generate a synchronized 12 MHz USB clock signal.

USB-compatible transceivers are provided for the upstream port and all downstream ports. Every downstream port supports both full- and low-speed connection by automatically setting the slew rate according to the speed of the device attached to the port.

Company Name: Texas Instruments

Address: Post Office Box 655303

Dallas, TX 75265

Telephone: 972-644-5580

Fax: 972-480-7800

Web: http://www.ti.com

Company Name: Thesys

Product Name: TH6503 - USB Low-Speed Interface

Features:

Supports up to two programmable endpoints for interrupt transfer

3 x 8 bytes FIFO

Data transfer at low speed

Supports suspend mode

Serial microcontroller interface

Register programmable

Programmable 1.5 MHz to 6 MHz out clock for microcontroller

Provides power supply for the microcontroller (3.3 volts or 5 volts)

Integrated oscillator for clock generation, supports 6 MHz quartz, ceramic resonator, or external clock input

Relatively simple external circuitry

Description: The TH6503 is an integrated circuit which enables the Universal Serial Bus (USB) to be connected to a microcontroller. The interface module contains all the components required to transmit data via the USB. The TH6503 has been developed for applications requiring a low speed interface to the USB. Any microcontroller can be used for control purposes. The TH6503 has been developed in conformity with USB Specification 1.0.

The TH6503 translates the data and control signals received from the USB in a serial format which can be read by the microcontroller. The data is stored in a FIFO buffer and can be called up from a standard microcontroller via a register programmable serial interface at any time and processed further. Data generated by peripherals is passed to the TH6503 with the same protocol and stored in a FIFO buffer until it is collected by the USB. The TH6503 translates all the data in the USB-specific format and generates the necessary control signals. The TH6503 requires a minimum number of external elements and can easily be implemented in a circuitry. It provides an external clock, which can be used to activate a microcontroller.

The TH6503 can be connected with any microcontroller via a serial interface. The serial outputs can be connected with any microcontroller port. The TH6503 clock out can be used to provide the clock pulse supply to the microcontroller.

Company Name:	Thesys
Address:	Haarbergstraße 61
	D-99097 Erfurt
	Germany
Telephone:	+49 361 427 8350

Fax: +49 361 427 6161
Email: detlef@thesys.de
Web: www.thesys.de

Company name: Winbond Electronics Corp.

Product Name: W81C180 USB Hub Controller

Description: W81C180 implements a medium speed (12Mhz) Universal Serial BUS (USB) hub controller. It supports four downstream ports and an I^{2}C serial interface to a microcontroller. The W81C180 is a compound USB device, which can be used to implement a USB hub with embedded function.

Company name: Winbond Electronics Corp.

Address: No. 4, Creation Rd. III

Science-Based Industrial Park

Hsinchu, Taiwan, R.O.C.

Telephone: 886-3-5770066, ext. 7018

Fax: 886-3-5792646

Email: khlin@winbond.com.tw

Web: http://www.winbond.com.tw

6.5 Power ICs

Company Name: Unitrode Corporation

Product Name: Universal Serial Bus Power Controller

Model Number: UCC3831

Features:

Fully USB compliant

Complete power control for monitor or other four-port self-powered hub

Four 5V current limited (>500mA) port outputs with individual enable

3.3V, 100mA local regulator for USB controller

Pre-regulator controller for regulation from monitor filament or other loosely regulated voltages

Available in 28 pin wide surface mount power package for excellent thermal performance

5V port can be also be used for 5V USB controller IC

Description: The UCC3831 Power Controller is designed to provide a self-powered USB hub with a local 3.3V regulated voltage as well as four 5V regulated voltages for USB ports. Each of the 5V output ports is individually enabled for optimal port control. Each port also provides an overcurrent fault signal indicating that the port has exceeded a 500mA current limit. The 3.3V linear regulator is used to provide power to the local USB microcontroller. This regulator is protected with a 100mA current limit and has a logic enable pin as well.

The UCC3831 can be configured to provide USB port power from a loosely regulated voltage such as a filament voltage internal to a monitor. Pre-regulation is provided by an internal linear regulator controller and one external N-channel MOSFET. The UCC3831 can also be configured without a using the pre-regulator stage by connecting the VREG pins to a regulated 5.5V 2A source.

The UCC3831 comes in a 28-pin wide SOIC power package optimized for power dissipation, and is protected by internal over-temperature shutdown mechanism, which disables the outputs if the internal junction temperature exceeds 150°C.

Since the UCC3831 uses linear regulators to provide port voltages, response to bus transients is excellent, and isolation from port to port is much better than a system where MOSFET switches are used. In a MOSFET switch controlled port, a short circuit or high current transient on one line will result in a droop in the 5V source, and may result in other ports falling out of regulation.

Design Tips: The UCC3831 can generate 3.3V from either an upstream 5V port or a local power supply anywhere from 4.4V to 9.0V. If an upstream port is used, the input capacitance should be limited to less than 10μF, per section 7.2.4.1 of the Universal Serial Bus Specification. Low cost aluminum electrolytic capacitors can be used for this function.

A pre-regulator stage is built into the UCC3831 to generate a regulated 5.5V from a loosely regulated higher voltage such as a monitor filament voltage. One external N-channel MOSFET is used to support this function. The 5.5V output is then connected into the 5V port regulators. This pre-regulator can be bypassed easily by connecting the VREG pins of the UCC3831 directly to a 5.5V source, such as a local power supply.

The 5V Port outputs should be decoupled with 120μF, per section 7.2.4.1 of the Universal Serial Bus Specification. No minimum load is required on these ports to maintain stability.

Unitrode Device Temperature Management and Thermal Characteristics design guidelines can aid in controlling the junction temperature of the UCC3831. Two square inches or more of copper should be used as a ground plane to provide a good thermal path for the UCC3831. Built-in thermal protection will protect the device in the event of an extended short circuit.

To prevent extended short circuit operation, the USB Hub controller should disable the individual port upon receiving an OCP indication.

Company Name:	Unitrode Corporation
Address:	7 Continental Boulevard
	Merrimack, NH 03054
Telephone:	(603) 429-8504
FAX:	(603) 429-8564
Email:	productinfo@unitrode.com
Web:	www.unitrode.com

Company Name: Micrel, Inc.

Product Name: USB High Side Power Switches

Part Numbers: MIC2525/MIC2526

Features:

Compliant to USB Specification 1.0

Low on-resistance, 100 mΩ typical

500mA minimum load current

Current limiting, 750mA typical

Thermal shutdown

Under-voltage lockout (UVLO)

Overcurrent flag output

Soft start circuit

Active-high and active-low enable

Flag output and enable compatible with 3.0v and 5.0V logic

Description: The MIC2525 and MIC2526 are single and dual high side switches optimized for USB power distribution. They are available with both active-high and active-low enables for flexibility in interfacing to any USB controller. They can be used in both USB self-powered and bus-powered hub applications. They are available in both 8-pin DIP and SO packages.

USB has very stringent requirements for power distribution. The MIC2525 and MIC2526 devices have been designed to solve all of the issues related to USB power distribution. The MIC2525/MIC2526 devices integrate a 100mΩ switch that meets USB requirements for voltage drop and regulation. In addition, these devices will automatically limit current to very safe levels (750 mA typical), which is well within the USB specification of five amps. Limiting current also helps preserve battery capacity in portable equipment like laptop PCs. Upon detecting a current limit situation, the MIC2525 and MIC25265 will provide a logic-compatible flag output to the local USB controller for use by the host software. These devices are also fully protected against shorts or other fault conditions due to a thermal shutdown circuit that monitors the temperature of the IC. Once the specified temperature limit is reached, the switch will shut down to prevent damage to the device. An undervoltage lockout (UVLO) circuit also prevents the switch from operating until the input voltage is present.

The MIC2525 and MIC2526 also feature a soft start circuit that provides slow and controlled turn-on of the switch. This feature limits the inrush current that occurs when charging the bulk capacitance of the devices attached to the downstream ports. By limiting this current, the USB voltage droop specification for the bus voltage is satisfied thereby insuring that downstream peripherals operate correctly.

Company Name: Micrel Semiconductor
Address: 1849 Fortune Dr.
San Jose, CA 95131
Telephone: 408-944-0800
Fax: 408-944-0970
Email: lasmaz@aol.com
Web: www.micrel.com

6.6 Other ICs

Company Name: Philips Semiconductor
Product Name: Analog Transceiver
Model Number: PDIUSBP11A
Features:

Supports 12 Mbit/s full speed and 1.5 Mbit/s low speed serial data transmission

Utilizes digital inputs and outputs to transmit and receive USB cable data

Compatible with the VHDL Serial Interface Engine from USB developers' conference

Complies with Universal Serial Bus Specification 1.0

Available in 14-pin, SO package

Description: The PDIUSBP11A is a generic USB analog transceiver front-end device. It is designed to allow direct interfacing to the USB Serial Interface Engine. Both 12 Mbit/s full speed and 1.5 Mbit/s low speed are supported. It is fully compliant to the USB specification 1.0. The PDIUSBP11A compliments the digital ASIC implementation of USB devices.

Company Name:	Philips Semiconductors
Address:	811 E. Arques Ave.
	Sunnyvale, CA 95066
Telephone:	408-991-3276
Fax:	408-991-2133
Email:	teresa.hardy@sv.sc.philips.com
Web:	www.semiconductors.philips.com

Company Name:	Philips Semiconductors
Product Name:	Stereo D/A Bitstream Converter
Model Number:	UDA1321

Features:

Complete stereo USB-DAC system with integrated filtering and line output drivers

Supports 12 Mbits/s full-speed serial data transmission mode

Supports USB-compliant audio multimedia devices over an industry-standard USB-compatible 4-wire cable

Fully automatic Hot Plug and Play operation

Supports multiple audio data I/O formats

On-chip timing reference recovery system including oscillator circuitry, using an external crystal for clock regeneration

Excellent audio performance: 90 dB dynamic range, 95 dB SNR

Operates from a 3.3 V power supply

Low power consumption

Housed in a small SOP28

Description: The UDA1321 is a stereo CMOS digital-to-analog bitstream converter. It extracts the encoded digital audio information, recovers the time reference lost by USB transmission, and passes the reconstructed digital audio signal through a high-quality digital-to-analog converter. In combination with a power amplifier IC, the UDA1321 performs all the signal processing required to construct a USB digital speaker.

Company Name:	Philips Semiconductors
Address:	811 E. Arques Ave.
	Sunnyvale, CA 95066
Telephone:	408-991-3276
Fax:	408-991-2133
Email:	teresa.hardy@sv.sc.philips.com
Web:	www.semiconductors.philips.com

Company Name:	Winbond Electronics Corp.
Product Name:	W81C380 USB Scanner Control Interface

Description: The W81C380 interfaces a scanner to the host computer via medium speed (12Mhz) USB. It supports DMA data transfer controller and standard microcontroller interface. With effective FIFO arrangement, the W81C380 supports high-speed data transfer to the host computer.

Other ICs

Company name:	Winbond Electronics Corp.
Address:	No. 4, Creation Rd. III
	Science-Based Industrial Park
	Hsinchu, Taiwan, R.O.C.
Telephone:	886-3-5770066, ext. 7018
Fax:	886-3-5792646
Email:	khlin@winbond.com.tw
Web:	http://www.winbond.com.tw

Company Name:	Lucent Technologies, Inc.
Product Name:	USS-720 Instant USB™, USB-to-Parallel Conversion IC

Features:

Device Features:

   3.3 V operation

   44-pin MQFP

   On-chip transceivers for USB

   Low power consumption allows part to be powered from USB connection

   Dual 192-byte, on-chip USB packet buffering for fast host response

   On-board nonvolatile storage for configuration-free power-up operation

   Fully compatible USB host device drivers available

   Fully compliant with current printer device class specification

IEEE-1284 (Parallel) Features:

Full support of compatibility, nibble, ECP, and EPP modes for true bi-directional communication

Hardware initiates and manages automatic negotiation for the fastest protocol available

Support of multiple logical channels

Throughput greater than 900 Kbytes/s (ECP mode)

Description: The USS-720 Instant USB integrated circuit provides a simple interface between universal serial bus (USB) and existing IEEE-1284 parallel port peripherals. It is suitable for integrated applications where the USS-720 is utilized inside the peripheral product. The IC also may be incorporated into USB-to-parallel adapter products.

Two USB channels are supported to allow logically concurrent communications between the USB, parallel port, and internal configuration.

Overview

The USS-720 presents one USB port and one IEEE-1284 enhanced parallel port to the outside world. Internally, the USS-720 contains an integrated USB transceiver, a universal device controller core, an IEEE-1284 core, storage for USB configuration data and data buffers, and control logic to tie the pieces together.

In use, the USB port is connected via a USB cable to a host computer, possibly with one or more USB hubs in between. Host software can send commands and data to the USS-720 and receive status and data from the USS-720 using the USB transport.

The IEEE-1284 enhanced parallel port is connected to a peripheral device. If the peripheral is IEEE-1284 capable, the associated features and communication modes may be used. The USS-720 directly supports four of the five IEEE-1284 modes: compatibility, nibble, EPP, and ECP. The fifth defined mode, byte mode, is not directly supported. Host software can manipulate the parallel port control and data lines manually in order to implement byte mode, IEEE-1284 extensions (e.g., 1284.3), or other, nonstandard communication protocols.

USB Port

The USB port on the USS-720 is electrically and logically compliant with the USB Specification 1.0.

Device Descriptor, Configurations, and Interfaces

The USS-720 has a default hard-coded configuration or can accept an override configuration from an external serial PROM. This configuration will be reported to the host during device enumeration via the get descriptor commands.

The default device configuration has one configuration (configuration 1) and one interface (interface 0). There are three alternate settings for the interface, and they include the following endpoints: (see Table)

Default Configuration

Interface	Alternate Setting	Endpoint Type	Endpoint Number
0	0	Control	0
		Bulk Out	1
0	1	Control	0
		Bulk Out	1
		Bulk In	2
0	2	Control	0
		Bulk Out	1
		Bulk In	2
		Interrupt	3

Pipes

Four pipes are defined: control, bulk in, bulk out, and interrupt.

Control Pipe

The control pipe is the default pipe, used for USB setup and control packets. Its maximum packet size is 8 bytes. The control pipe is also used for class- and vendor-specific commands which:

    Configure class- and vendor-specific features

    Read/Write the parallel port registers

    Read/Write an address byte from/to the peripheral in EPP mode

Chapter 6: Examples of USB Hub and Device ICs

Read/Write a data byte from/to the peripheral in EPP mode (but multiple bytes can be transferred more efficiently via the bulk pipes).

Class- and vendor-specific commands consist of a USB command packet followed by a USB OUT packet.

Bulk In Pipe

The bulk in pipe is used to read data bytes from the peripheral in nibble, ECP, and (with care) EPP mode. Its maximum packet size is 64 bytes. In EPP mode, there is no way for the USS-720 hardware to detect the end of valid data from the peripheral. So the host must first write the desired data length to the EPP data length register. That triggers the USS-720 to begin reading bytes from the peripheral, and those bytes are returned to the host in the subsequent read from the bulk in pipe. (If only a few bytes are being transferred it is more efficient for the host to write each byte to the EPP data register, which transfers the byte in a single operation). Compatibility mode is forward only (host to peripheral), so an attempt to read from the bulk in pipe while in compatibility mode will always result in a NAK response to the host.

Bulk Out Pipe

The bulk out pipe is used to send data to the peripheral in compatibility, EPP, or ECP mode. Its maximum packet size is 64 bytes. Nibble mode is reverse only (peripheral to host). If the host sends data to the bulk out pipe while in nibble mode the USS-720 will automatically negotiate into compatibility mode, send the data, and negotiate back into nibble mode without further host interaction.

Interrupt Pipe

The interrupt pipe is used to report changes in parallel port status to the host. It is two bytes in length, as required by the USB specification. When the interrupt pipe is enabled by host software, the USB hardware on the host automatically polls the USS-720 periodically. The USS-720 returns two bytes of status whenever the parallel port status has changed since the last poll, and returns nothing otherwise. The two bytes are the parallel port status register and a delta byte containing buffer status information and a flag indicating when the nACK parallel port signal has changed. (nACK is the signal that causes a hardware interrupt in a conventional parallel port hardware.) This enables the host to detect and react to parallel port and buffer status changes without polling the control endpoint.

Inter-Pipe Synchronization

With commands and data going to different pipes, and data potentially being buffered inside the USS-720, it could be difficult for host software to maintain serialization of operations on the peripheral. The interrupt pipe status mechanism described above can be used to alleviate this problem. Software can use the port status and buffer status information thus returned to determine when buffered data has been sent and when port control commands have been processed and it is safe to continue. Since the information is returned to the software automatically and only when it changes, overhead for the host operating system and driver software is kept low.

IEEE-1284 Port

Overview

The IEEE-1284 port on the USS-720 is compliant with the IEEE-1284 1994 standard. Compatibility, nibble, EPP, and ECP modes are directly supported in hardware, and a Microsoft standard register interface is also available to the host. Status bits in the standard registers are used to indicate analogous conditions in the USS-720; for example, the FIFO empty and full status bits are used to indicate USB packet buffer empty and full. Hardware assist is included for the most commonly performed or most tedious (for software) functions. The intent is to allow existing IEEE-1284 software and methods to be ported to the USS-720 quickly while improving performance and reducing software overhead.

The port is capable of handshaking data to and from the peripheral. Thus the host software can send the data without having to manually toggle control lines or monitor status bits.

The port has auto-negotiation capability, meaning that it can negotiate with the peripheral to switch from one IEEE-1284 mode to another automatically, without host interaction. The host can allow the hardware to handle negotiation completely and automatically, in which case the USS-720 will automatically transfer bi-directional data in the fastest mode supported by the peripheral, negotiating between modes when necessary. Alternatively, the host may prefer to use semiautomatic negotiation in which the host manually negotiates into an IEEE-1284 mode and then sets the USS-720 into that mode. Then the host can perform bi-directional data transfers and the hardware will automatically switch into forward or reverse as required. Again, no host interaction is required to perform the switch.

Registers

Nine parallel port registers are available to the host. They are read and written via the control pipe.

Parallel Port Registers

	Address	Access	Function
Data	0	Read/Write	Read/Write the parallel port data lines.
Status	1	Read	Read the state of the parallel port status lines.
Control	2	Write	Write the state of the parallel port control lines.
EPP/ECP Address	3	Read/Write	Read/Write an address byte in EPP or ECP mode.
EPP/ECP Data	4	Read/Write	Read/Write a data byte in EPP or ECP mode.
EPP Data Length	5	Read/Write	Read/Write the number of bytes for a subsequent bulk in read in EPP mode. (0 means 256 bytes.)
ECP Cfg A	6	Read	Constant value, 0x10.
ECP Cfg B	7	Read	Constant value, 0x80.
ECP ECR	8	Read/Write	ECP extended control register.
Register	Address	Access	Function
Data	0	Read/Write	Read/Write the parallel port data lines.
Status	1	Read	Read the state of the parallel port status lines.
Control	2	Write	Write the state of the parallel port control lines.
EPP/ECP Address	3	Read/Write	Read/Write an address byte in EPP or ECP mode.
EPP/ECP Data	4	Read/Write	Read/Write a data byte in EPP

			or ECP mode.
EPP Data Length	5	Read/Write	Read/Write the number of bytes for a subsequent bulk in read in EPP mode. (0 means 256 bytes.)
ECP Cfg A	6	Read	Constant value, 0x10.
ECP Cfg B	7	Read	Constant value, 0x80.
ECP ECR	8	Read/Write	ECP extended control register.

External Circuitry Requirements

The USS-720 is intended to be a single-chip solution. As such, the USB transceiver and the IEEE-1284 drivers have been integrated on-chip. The only external requirements are a 3.3 V supply and a 1.5 kΩ pull-up resistor for the D+ pin. If the internal oscillator is used, a 12 MHz crystal along with bias capacitors, and a 48 MHz or 12 MHz clock signal if the internal oscillator is not used. A 5 V supply and USB and/or IEEE-1284 connectors might also be needed, depending on the application.

Electrical Specifications

The USS-720 is a 3.3 V part, and has a separate pin for power to the IEEE-1284 drivers.

Company Name: Lucent Technologies, Inc.

Microelectronics Group

Address: 555 Union Boulevard

Room 30L-15P-BA

Allentown, PA 18103

Telephone: 1-800-372-2447

Fax: 610-712-4106

Email: docmaster@micro.lucent.com

Web: http://www.lucent.com/micro

7. USB Development Tools and Helpful Sources

7.1 Purpose of this Chapter

This chapter lists sources of useful information and development tools. The development tools include both PC-based products and stand-alone units. For the most part, they have three key components: a signal generator, a decoder, and a monitor to read the packets generated.

Also listed are organizations which provide design and consulting services for hardware, software (including drivers), silicon, mechanical, custom cables, custom connectors, seminars and training as a quick reference. A complete list of USB vendors is included in Chapter 8.

Bus Analyzer

    FuturePlus, p. 175

Custom Device Drivers

    Ashley Laurent, p. 178

Hardware Emulator

    Cypress Semiconductor, p. 180

Input Device Developer's Kit

    USAR Systems, p. 180

Host Controller Developer Kit

    CMD, p. 181

Monitor firmware

    SystemSoft, p. 183

Purpose of this Chapter

OS Utility

>Phoenix Technologies, p. 187

Overcurrent Protection

>Raychem, p. 190

Printer firmware

>SystemSoft, p. 191

Product Development Services

>e^{TEK} Labs, p. 195

Protocol Analyzer

>Genoa Technology, p. 195

Silicon design services

>MicTron, p. 197

Simulation Models

>Phoenix, p. 198

>Sand, p. 201

Synthesizable Cores

>Decicon, p. 204

>Oki, p. 206

>Phoenix, p. 216

>Sand, p. 223

Transceiver Design

>Decicon, p. 231

Workshops, conferences, and books

>Annabooks, p. 232

The following information was received in response to inquiries sent to the USB community during the preparation of this book. The author apologizes to any vendors who may have submitted information that is not included here; material in incorrect format,

Chapter 7: USB Development Tools and Helpful Sources

lost material, or unreadable media may have been at fault. In any case, the author or publisher cannot be responsible for omissions, inaccuracies, or errors in the published information. The following information is intended as a guide to the reader as to the types and varieties of USB products becoming available. Please also see the list of vendors in Chapter 8.

7.2 Bus Analyzer

Company Name: FuturePlus Systems

Product Name: USB Bus Analyzer

Model Number: FSUSB

Features:

Passive USB bus analysis

Complete USB serial to parallel decode

Automatic detection and operation at high-speed (12Mbits/s) or low-speed (1.2Mbits/s), including dynamic speed changing

Automatic USB reset detection

Address and endpoint specified in token packet held until transfer completes

Allows for easy triggering, store qualification and performance monitoring of specific endpoints

Supports the full USB specification

Supports all types of data transfers, including isochronous transfers

Supports dynamic hot swapping

Uses your existing HP logic analyzer; requires only two pods

Complete configuration files and USB Transaction Inverse Assembler supplied for your HP logic analyzer

Uses HP's enhanced triggering capabilities, cross-domain analysis, store qualifiers, and system performance software for complete USB performance monitoring

Description: The USB preprocessor provides two functions:

1) Provides an electrical and mechanical interface from the Universal Serial Bus to Hewlett-Packard logic analyzers for passive bus analysis.

2) Provides test points to measure the power and signal fidelity of the USB bus.

State Analysis Mode

The software included with the FSUSB contains complete configuration files and a FuturePlus USB Transaction Inverse Assembler for your HP logic analyzer. In State Analysis mode, the analyzer master clock is derived from the USB Protocol. The USB serial data is converted to parallel data in the preprocessor regardless of high or low speed operation. Any USB resets are automatically detected. Since the address and endpoint specified in the token packet are held until the transfer is complete, triggering, store qualification and performance monitoring of specific endpoints is easy!

The enhanced triggering capabilities of your HP logic analyzer allow you to trigger on (1) any address and endpoint, (2) any data pattern, (3) any data CRC, (4) any USB error (CRC fail, serial bit stuff error, and missing frames), (5) bad or invalid PID's, or any combination of these.

Store qualifiers allow the user to store any combination of (1) any address and endpoint, (2) any data pattern, (3) any data CRC, (4) any PID type, or any combination of these.

All USB cycles and transaction identifiers (SOF, OUT, IN, SETUP, DATA0, DATA1, ACK, NAK, STALL, and PRE) are decoded by protocol-sensitive clocking logic and presented as separate bits to the logic analyzer. These packet identifiers will allow the user to (1) store all USB traffic, (2) store only certain packet types, (3) store only packets to and from a certain function. The FuturePlus Transaction Inverse Assembler makes analyzing the resulting

Chapter 7: USB Development Tools and Helpful Sources

stored USB traffic easy and accurate. Another good feature: the electrical power for the FSUSB circuitry is drawn from the logic analyzer, not your USB bus!

Timing Analysis Mode

The USB Analyzer has a third pod dedicated to timing analysis of the USB serial bit stream. The FSUSB in timing mode provides a unique state by state view of the USB serial interface engine (SIE). This mode allows for:

- Shadowing the state of the target USB SIE when that SIE state is unavailable

- Comparing the state of the target USB SIE with that of the FSUSB SIE

- Making accurate timing measurements of USB events

- Accurate USB protocol violation detection

- Accurate USB signaling violation detection.

Signals on Pod 3

Signal Name	Description
CLK 12	Recovered Clock
MDATA	Recovered Serial Data
SOFTIC	Start of Frame. 1 ms timer generated from recovered start of frame
EOP2_0	End of Packet state machine
LBC3_0	Load Byte Count State Machine
RST2_0	USB Reset State Machine
FEOPR	End of Packet
FEOSYN	End of Sync
LSDET	Low Speed Detect

Cross-Domain Analysis

If you analyzing data in multiple domains, use this preprocessor to monitor the USB, and then use another FuturePlus Systems preprocessor to monitor your other bus. Preprocessors are available for the PCI, ISA, VME, VXI, PMC, and SIMM buses. You can create your own custom measurement system, cross-domain trigger between buses, and view data from multiple buses simultaneously in the same display. In a similar fashion, you could connect a

Custom Device Drivers

preprocessor for your host processor to another logic analyzer card. You could then use HP's Software Analyzer (B4620A) to view source code, code execution, and the corresponding USB packet transfers simultaneously.

HP Logic Analyzers Supported

Two logic analyzer pods are required; three pods also allow full timing analysis.

HP 16500B/C Logic Analysis System modules:

HP 16505A Prototype Analyzer

HP 16554A, 16555A, and 16556A (4 pods)

HP 16550A (6 pods)

HP Benchtop Logic Analyzers:

HP 1660A/C, 1661A/C, 1662A/C, (8,6,4 pods)

HP 1663A/C, 1664A (2 pods)

HP 1670A/D, 1671A/D, 1672A/D (8,6,4 pods)

Company Name:	FuturePlus Systems
Address:	3550 North Academy Blvd.
	Colorado Springs, CO 80917
Telephone:	719-380-7321
Fax:	719-380-7362
Email:	102416.3136@compuserve.com
Web:	http://www.futureplus.com

7.3 Custom Device Drivers

Company Name: Ashley Laurent, Inc.

Description: Custom device drivers and system software for Microsoft operating systems. Experience includes writing all major NDIS drivers, USB, SCSI/IDE and mass storage, 2-D/3-D graphics, IFS (Installable File System), and telephony/modem drivers. Licensing is available for NDIS VPN (Virtual Private Networking) IP tunneling drivers. On site or fixed price.

Have written USB drivers for the USS-720 USB<->1284 device and is currently engaged in projects involving HID, Composite Devices, and Still Image Drivers for Digital Camera and Scanner. Especially helpful to USB manufacturers is the ability of Company to adapt drivers designed for legacy interfaces such as parallel and serial, to the USB driver model.

The company's wide range of experience includes network, mass storage, and graphics devices.

Specialization:

Disk Drives (SCSI/IDE)

HAL

IEEE 1394

Imaging

Networking

Sound

Video

USB

WDM

Data Acquisition, Medical, and Industrial Applications

Certified Microsoft Solution Provider

Company Name:	Ashley Laurent, Inc.
Address:	707 West Avenue, Suite 201
	Austin, TX 78701
Phone:	(512) 322-0676
Fax:	(512) 322-0680
E-mail:	salesbox@osgroup.com

Web: http://www.osgroup.com

7.4 Hardware Emulator

Company Name: Cypress Semiconductor
Product Name: USB Developer's Kit
Model Number: CY365x

Description: The CY365x is a full-speed hardware emulator to help you develop firmware and system drivers for Cypress USB microcontrollers. The kit contains assembly software, debug software, source code, and complete documentation.

Company Name: Cypress Semiconductor
Address: 3901 North First Street
San Jose, CA 95134
Telephone: 1-800-858-1810
Fax: 408-943-6848
Web: http://www.cypress.com

7.5 Input Device Developer's Kit

Company Name: USAR Systems
Product Name: USB Developer's Kits for Input Devices
Model Numbers: EVK5-USB-100, EVK7-UNI-USB-100, and EVK7-USBM-100

Features:

Available for keyboards, mice and touch screens

Includes all cables, boards, documentation, software and appropriate input device

Ideal for peripheral and driver development and system development and testing

Includes special upgrade board

Can always remain compliant to the latest specification revision

Combination touch screen and HulaPoint Developer's Kit available

Description: USAR Systems is offering a series of USB HID Developer's Kits to aid OEMs in the design of new peripherals, drivers and USB compatible systems. The kits, available for keyboards, mice, and touch screens, contain everything a designer needs to create a device fully compliant to the latest USB specification revision, including function IC/USB HID core code, board, cables, documentation, and testing software. The kits also contain a special upgrade board. When a change in the USB specification occurs, USAR will update its code. Designers may then use the upgrade board to download the latest revision directly to their kits. On-going compliance with the USB specification is thus assured.

Company Name:	USAR Systems
Address:	568 Broadway, #405
	New York, NY 10012
Telephone:	212-226-2042
Fax:	212-226-3215
Email:	egooch@usar.com

7.6 Host Controller Developer's Kit

Company Name:	CMD Technology
Product Name:	Host Controller Development Kit
Model Number:	CSA-6700-DK
Features:	

A PCI to USB add-in card that is OpenHCI compliant (CMD's CSA-6700)

Beta copies of CMD's Windows 95 and DOS Host Controller Drivers

CMD's API specifications for device driver development

Description: By developing device drivers for the CMD APIs, peripheral manufacturers can immediately access the huge installed base of PCI computers that do not have USB capabilities.

The USB Development Kit enables manufacturers of USB devices to sell those devices to the existing installed base of PCI computers that currently do not support USB. Because users of those systems can upgrade to USB immediately when the add-in card is bundled with the USB device, makers of those devices don't need to wait for the USB installed base to be created. In addition, with CMD's host controller drivers, there is no need to wait for USB support in the operating system.

The USB Development Kit includes CMD's CSA-6700 PCI to USB add-in card which plugs seamlessly into any PCI 2.1 compliant PC motherboard for immediate upgrade to USB. Also included is a complete Windows 95 development environment for VxD based device drivers.

CMD also supports a complete DOS development environment for USB devices operating under a DOS environment.

The CSA-6700-DK USB development kit provides instant access to a development environment for USB. By bundling CMD's USB add-in card and host controller driver with your peripheral, there is no need to wait for the USB installed base to mature before entering the market. All that's required is completion of the VxD or DOS based device driver.

The CSA-6700-DK USB development kit is being sold for development purposes only at this time. Anyone may order the kit; however, CMD will provide technical support only to those companies that apply and qualify for a beta ID. To qualify for a beta ID, a company must be actively developing device drivers with the kit and have plans to ship the CSA-6700 in volume.

Company Name: CMD Technology, Inc.
Address: 1 Vanderbilt
Irvine, CA 92618
Telephone: 800-426-3832
714-454-0800
Fax: 714-455-1656
Email: info@cmd.com
Web: www.cmd.com

7.7 Monitor Firmware

Company Name: SystemSoft Corporation
Product Name: SystemSoft USB Monitor Suite
Features:

Provides monitor control through USB

USB monitor class-compatible

Designed for easy OEM customization

Based on SystemSoft's USB interface firmware

Link-in integrated hub support

Includes all required host-side software

Scaleable Plug and Play

Win32 Driver Model (WDM) for Windows 95 and Windows NT

More cost-effective than doing it yourself

Ensures compliance with USB standards

Description: SystemSoft's USB Monitor Suite provides immediate Universal Serial Bus compatibility. This product offers everything you need to provide host-based, on- screen control of your USB-capable monitor. The device-side firmware is completely compatible with the USB Monitor Class specification. It's based on

SystemSoft's USB Interface Firmware (UIF), delivering hardware-independent USB access for any application. Integrated hub functionality is just a matter of linking with the right UIF library. On the host-side you get the Win32 Driver Model (WDM) support you need. SystemSoft also offers the 32-bit dynamic link library (DLL) and applications software for a complete self-configuring solution.

Provides Monitor Control Through USB

Monitors offer a wide range of possible controls. From the common brightness, contrast, size, and position controls all monitors support, to more exotic controls like moire and linearity. Until now, most manufacturers have provided on-screen displays using controls local to the monitor. These displays have been limited by the processor power and program memory available on the monitor. With USB, the control task and user interface management are moved to the host system, with all of its horsepower to create impressive control programs and test patterns. SystemSoft's Monitor Wizard automatically adapts to the capabilities of the attached monitor. As host systems become more complex and offer multiple monitor configurations, SystemSoft's Monitor Wizard will support them all.

USB Monitor Control Class-Compatible

SystemSoft has been working with the USB Monitor Working Group to develop the USB Monitor Control Class specification. Focused on the unique concerns of USB connectivity, this standard ensures your device will work with a wide range of host systems running USB-compatible operating systems. With SystemSoft's Monitor Wizard, you get the necessary host-based software to take advantage of your USB Monitor Control device.

Designed for Easy OEM Customization

You need product differentiation to survive in the PC marketplace -- after you deliver compatibility. That's why SystemSoft has designed its USB Monitor Software Suite to enable you to customize our compatible solutions. On the device side, it offers support for a number of interfaces to monitor control hardware including RS-232 and I^{2}C. On the host-side, your device is

Chapter 7: USB Development Tools and Helpful Sources

recognized and installed automatically by a class-specific WDM driver. Internationalization of the application program is available as an install or run-time option.

Based on SystemSoft's USB Interface Firmware (UIF)

SystemSoft has invested multiple man-years developing an abstraction of USB connectivity. The result is a firmware layer providing a hardware-independent programming interface for USB hardware. Our USB Interface Firmware (UIF) allows high-level digital data stream management independent of the underlying USB hardware. All control, bulk, interrupt, and isochronous data transfers are completely handled by the UIF. Higher-level device firmware, like our Monitor-specific code, just passes buffers (empty or full) to the UIF. Device configuration is automatic based on descriptor information coded into the firmware at compile-time. Standard USB requests are also handled completely by the UIF. That means SystemSoft USB solutions are portable, easy to apply to various opportunities, and are based on interface management code that has had the benefit of testing in a wide variety of situations.

Link-In Integrated Hub Support

Since all SystemSoft USB designs are based on the same USB Interface Firmware interface, adding integrated hub support is as simple as linking with another object library. For example, if you've developed to our UIF (or are using a product incorporating our UIF like the Monitor Control Support) switching from an Intel 8x930 Ax (without a hub) to an Intel 8x930 Hx (with a hub) just requires a different version of the UIF.

Includes All Required Host-Side Software

SystemSoft provides complete end-to-end solutions, including all of the required host-side software. For our Monitor Control Support, the Win32 Driver Model (WDM) Monitor Control Driver is included. The same driver runs under Windows 95 and Windows NT. It provides a simple IOCTL and asynchronous notification interface to our dynamic link library (DLL) and Monitor Wizard application.

The Monitor Wizard automatically configures the user interfaces for the controls provided by the monitor. Test patterns are provided to assist in monitor adjustment. Unavailable controls are disabled, providing a consistent and easy to understand user interface.

WDM Driver for Windows 95 and NT

With the release of the USB Supplement for OEM Service Release 2 (OSR2) of Windows 95, Microsoft has defined the future of Windows drivers -- the Win32 Driver Model. It combines the best of Window NT Kernel Mode Drivers with the Plug and Play of Windows 95. You can write one driver to work in both Windows 95 and NT. SystemSoft's solutions include WDM drivers tested in both environments and, where possible, application code is developed to run in both environments.

More cost-effective than doing it yourself

SystemSoft's USB Monitor Suite utilizes standard software components that are easily tailored for specific types of port peripheral devices. It leverages a large body of code that has been tested and used in multiple applications.

SystemSoft employs software developers who are experts in implementing USB solutions. For hardware-oriented customers, this limits the amount of time required to learn USB, and allows companies to keep focused on their core competencies. It also limits the engineering mistakes that may result from not understanding the USB Standard. Once development is complete, a USB Hardware Integration Lab is available for testing your hardware.

Ensures compliance with USB standards

SystemSoft's USB Development Team knows the USB specifications and the vision behind them. The Company has been involved with USB from the beginning. Key SystemSoft employees wrote Chapters 9 (Device Framework) and 10 (USB Hosts: Hardware and Software) of the USB Specification, co-authored the Common Class Specification, and currently serve as Chair of the USB Device Working Group. The Company also enjoys a close working relationship with Intel and Microsoft.

Company Name: SystemSoft Corporation

Address: 2 Vision Drive

Chapter 7: USB Development Tools and Helpful Sources

	Natick, MA 01760
Telephone:	(508) 651-0088
Fax:	(508) 651-8188
Email:	usb@systemsoft.com
Web:	www.systemsoft.com/products/usb

7.8 OS Utility

Company Name:	Phoenix Technologies
Product Name:	USBWorks

Features:

Essential Windows 95 utility software for PC97

Universal Serial Bus Plug and Play made easier

Auto-launches the correct application when plugging in a USB peripheral

Monitors power on USB to detect device shut-down

Monitors bandwidth on USB to detect overload

Displays status on all USB peripherals connected

Controls and supports any USB peripheral

Extendable and upgradeable device support

Wizard for USB device configuration

OEM customizable

Description:

Universal Serial Bus Made Easier

USBWorks, a Windows 95 application, enhances the basic operating system support for USB so that adding new peripherals to PCs with USB is as easy as plugging in a connector. This utility helps peripheral and PC manufacturers put end users in control of USB, instead of calling for technical support. When bundled with a

OS Utility

new PC or peripheral, USBWorks addresses specific device configuration requirements and provides system-differentiating features to the end user.

USB — A Complex Standard

PCs shipping with USB can be very difficult to configure. For instance, the Universal Serial Bus is limited to 12 Mb/s bandwidth and 500 mA of current, yet it can allow up to 127 devices to be connected at the same time. This means a USB enabled PC can run out of bandwidth or use up power on the bus very quickly. This creates a confusing tangle for end users to unravel. OEMs must now support unprecedented levels of cross-compatibility as users mix different devices, including speakers, modems, cameras, printers, and game input devices on the same bus. USBWorks simplifies management of USB devices for the end user.

Better Control of USB Devices

USBWorks is a sophisticated Win32 COM-based application that complements the standard Windows Device Manager. It collects and displays information on all operational USB devices, even if they are not properly configured. It helps the user identify and resolve invalid connections, overcurrent, and out-of-bandwidth problems. It even allows applications to automatically launch when specific USB devices are attached.

User-Friendly Display of USB Ports and Peripherals

The USBWorks visual interface allows users to easily comprehend USB bus topology, device configuration, device bandwidth, and device power utilization. USBWorks includes the ability to name and personalize devices, much like volume names for hard disks. The visual interface is also used to determine USB port availability and identify devices recognized by the operating system. The visual interface also includes extensions for OEMs to add OEM-specific configuration or application pages.

Real-Time Monitoring of USB Power and Bandwidth

USBWorks monitors the USB device tree and determines the USB bus power and frame-based data bandwidth requirements. It reports bandwidth allocation limits and explains the consequences.

For example, if bandwidth or bus power consumption exceeds the maximum allowed, USBWorks will report the problem to the end user and recommend appropriate action to fix it.

Automatically Launches USB Peripheral Applications

USBWorks allows users to assign applications to launch when specific USB devices are plugged in. USB devices have unique identifiers, including manufacturer and device type tags. USBWorks allows users to associate these tags with specific applications that are launched when devices are recognized. For example, plugging in a modem may automatically trigger an Internet login, or plugging in a specific game device may automatically launch a particular game.

OEM Customizable Device Support

USBWorks supports OEM-specific configuration pages, allowing OEMs to customize the application to specific peripheral needs. For example, a moving camera application may utilize an additional configuration page to control focus, panning, or even display moving images – features not native to the operating system. USBWorks will continue collecting OEM-specific pages as new devices are added. For example, if a modem added later to the system with the camera, a modem page containing ISDN or POTS configuration settings would be added to USBWorks.

Phoenix: A Leader in USB Support

Phoenix Technologies is the world's largest supplier of compatibility software to the PC industry and is implementing the USB compatibility standard in millions of next generation PCs. Phoenix's USB technology also includes PhoenixBIOS™ USB BIOS Extensions, and Virtual Chips™ host, hub, and function synthesizable cores for USB-compatible silicon designs. Phoenix is the author of several USB standard specifications, such as the USB PC Legacy Device Class Specification, the USB Printer Class Specification, and the USB Mass Storage Class Specification.

Company Name: Phoenix Technologies, Ltd.

Address: 411 E. Plumeria

San Jose, CA 95134

Telephone: 408-570-1000
Fax: 408-570-1001
Web: www.phoenix.com

7.9 Overcurrent Protection

Company Name: Raychem Corporation
Product Name: PolySwitch Resettable Fuse
Features:
  Conformity with USB Specification, Revision 1.0
  Compliace with:
    US1950/IEC950 safety requirements
    Windows 95 and PC97/98 standards
  Resettable protection
  UL-recognized safety device
  Low cost

Description: PolySwitch resettable fuses are polymeric positive temperature coefficient (PTC) devices appropriate for overcurrent protection in self-powered and bus-powered USB applications. They are suitable for both ganged port protection and individual port protection.

Company Name: Raychem Corporation
Address: 300 Constitution Dr.
                     Menlo Park, CA 94025
Telephone: (800) 227-7040
Web: www.raychem.com

7.10 Printer Firmware

Company Name: SystemSoft Corporation

Product Name: SystemSoft USB Printer (Parallel Port Adapter) Suite

Features:

Converts printers and other parallel port devices to USB

USB Printer Class Compatible

Designed for easy OEM customization

Based on SystemSoft's USB interface firmware

Link-in integrated hub support

Includes all required host-side software

Scaleable Plug and Play

Win32 Driver Model (WDM) for Windows 95 and Windows NT

More cost-effective than doing it yourself

Ensures compliance with USB standards

Description: The SystemSoft USB Printer (Parallel Port Adapter) Suite brings you Universal Serial Bus compatibility immediately. SystemSoft offers everything you need to convert your existing printer or other parallel port device to take advantage of the tremendous Plug and Play connectivity offered by USB. Device-side firmware is completely compatible with the USB Printer Class specification. Best of all, it's based on SystemSoft's USB Interface Firmware, delivering USB-controller-independent USB access for any application. On the host-side you get the Win32 Driver Model (WDM) support you need. USB printer solutions include an enhanced Print Monitor to connect to the Windows printer stack.

Converts Printers and other Parallel Port Devices to USB

The SystemSoft USB Printer Suite takes a 1284 parallel port and mates it with USB. Now you can convert your parallel interface devices to USB with minimal effort and in minimum time. Enjoy the huge demand for USB peripherals. No more parallel port hassles. USB allows up to 127 devices to be connected to a single system, all starting with one port on a host. And USB loads any required host-side drivers automatically when your device is connected. It's true Plug and Play.

USB Printer Class Device Compatible

SystemSoft has been working with the USB Printer Working Group to develop the USB Printer Class specification. Focused on the unique concerns of USB connectivity, this standard ensures your device will work with a wide range of host systems running USB compatible operating systems. In most cases you will continue to use the same printer device drivers you've been using. USB appears to them to be just another port like LPT1 or COM2.

Designed for Easy OEM Customization

You need product differentiation to survive in the PC marketplace -- after you deliver compatibility. That's why SystemSoft has designed the USB Printer Suite to enable you to customize it to your specifications. On the device side, it offers support for uni-directional parallel port hardware, or bi-directional interfaces using nibble, EPP or ECP. On the host-side, your device is recognized and installed, automatically loading the right WDM driver. Internationalization is available as an install or run-time option.

Based on SystemSoft's USB Interface Firmware (UIF)

SystemSoft has made a significant investment in developing an abstraction of USB connectivity. The result is a firmware layer that provides a hardware independent programming interface for USB hardware. SystemSoft's USB Interface Firmware (UIF) allows high-level digital data stream management independent of the underlying USB hardware.

All control, bulk, interrupt and isochronous data transfers are completely handled by the UIF. Higher-level device firmware, like the Printer (Parallel Port Adapter) -specific code, just passes buffers (empty or full) to the UIF. Device configuration is automatic based on descriptor information coded into the firmware at compile-time. Standard USB requests are also handled completely by the UIF.

Chapter 7: USB Development Tools and Helpful Sources

That means SystemSoft USB solutions are portable, easy to apply to various opportunities, and based on interface management code that has had the benefit of testing in a wide variety of situations.

Link-In Integrated Hub Support

Since all SystemSoft USB designs are based on the same USB Interface Firmware API, adding integrated hub support is as simple as linking with another object library. For example, if you've developed to SystemSoft's UIF (or are using a product incorporating the UIF like a Printer), switching from an Intel 8x930 Ax (without a hub) to an Intel 8x930 Hx (with a hub) only requires a different version of the UIF.

Includes All Required Host-Side Software

SystemSoft provides complete end-to-end solutions, including not only device firmware, but all of the required host-side software, as well. For our Printer (Parallel Port Adapter) Suite, the Win32 Driver Model (WDM) Printer Driver is included. The same driver runs under Windows 95 and Windows NT, providing a simple read/write interface for digital data streams with parallel port status read from the device using a USB Printer Class-Specific request.

Scaleable Plug and Play

Our host-based software is designed from the beginning for multiple device support. With USB, you never know how many devices of the same type might be attached. How do you name attached devices, find them, and connect to them? With USB the question is not how do you do it, but how do you do it so it's compatible and scaleable. Our solutions address these issues.

WDM Driver for Windows 95 and NT

With the USB Supplement for OEM Service Release 2 of Windows 95, Microsoft has defined the future of Windows drivers. Known as the Win32 Driver Model, it combines the best of Window NT Kernel Mode Drivers with the Plug and Play of Windows 95. Now you can write one driver to work in both Windows 95 and NT. SystemSoft's solutions include WDM drivers tested in both environments and, where possible, application code is developed to run in both environments.

More cost-effective than doing it yourself

SystemSoft's USB Printer (Parallel Port Adapter) Suite utilizes standard software components that are easily tailored for specific types of parallel port peripheral devices. It leverages a large body of code that has been tested and used in multiple applications.

SystemSoft employs software developers who are experts in implementing USB solutions. For hardware-oriented customers, this limits the amount of time required to learn USB, and allows companies to keep focused on their core competencies. It also limits the engineering mistakes that may result from not understanding the USB Standard. Once development is complete, a USB Hardware Integration Lab is available for testing your hardware.

Ensures compliance with USB standards

SystemSoft's USB Development Team knows the USB specifications and the vision behind them. The Company has been involved with USB from the beginning. Key SystemSoft employees wrote Chapters 9 (Device Framework) and 10 (USB Hosts: Hardware and Software) of the USB Specification, co-authored the Common Class Specification, and currently serve as Chair of the USB Device Working Group. The Company also enjoys a close working relationship with Intel and Microsoft.

Company Name: SystemSoft Corporation

Address: 2 Vision Drive

Natick, MA 01760

Telephone: (508) 651-0088

Fax: (508) 651-8188

Email: usb@systemsoft.com

Web: www.systemsoft.com/products/usb

Chapter 7: USB Development Tools and Helpful Sources

7.11 Product Development Services

Company Name: e^{TEK} Labs

Product Name: USB product development services

Description: e^{TEK} Labs software and hardware development services are composed of a combination of consulting services, operating system drivers, peripheral firmware, and ASIC development as a foundation for OEMs to build upon and complete their own USB products and system designs. The ultimate goal is to quickly integrate USB technology into existing products and provide assistance in the development of entirely new products, some of which would not have been possible without the introduction of USB.

An active member of the USB Implementers Forum, e^{TEK} Labs has expertise in device driver development for DOS, Windows, Windows 95, UNIX, and OS/2 environments. e^{TEK} Labs has been providing high quality device product development services for technology companies for more than 10 years.

Company Name:	e^{TEK} Labs
Address:	1057 East Henrietta Rd.
	Rochester, NY 14623
Telephone:	716 292-6400
Fax:	716 292-6273
Email:	Ptravers@eteklabs.com
Web:	www.eTekLabs.Com

7.12 Protocol Analyzer

Company Name:	Genoa Technology
Product Name:	USB Protocol Analyzer
Features:	

195

Protocol Analyzer

Captures and decodes all USB events including standard requests

Flexible triggering maximizes data capture

Consolidates data from multiple pipes

Filtering options greatly enhance debugging

Description: The USB Protocol Analyzer from Genoa Technology greatly simplifies the task of isolating USB protocol problems by making it easy to capture and analyze USB traffic at any branch in a USB system, from signal layer through data layer. The USB Protocol Analyzer works in conjunction with an HP 16500B/C mainframe w/ 16555A logic analyzer card, or optional Genoa GPU data acquisition board. Setup of the HP Logic Analyzer is accomplished through the USB Protocol Analyzer, which provides nine data capture options in addition to those already provided by the logic analyzer. These setup options greatly enhance the logic analyzer's ability to capture USB traffic, by triggering capture according to the occurrence of USB events. Capture features include 5x oversampling (five times the bus clock speed or 1/5 of full-speed bit time) and time stamping of all bus transactions even during idle periods. Both 12 Mbit and 1.5 Mbit signaling rates are supported

With the USB Protocol Analyzer, you can easily switch views between events, traffic, and data. Because the position in time is preserved, it is easy to see the relationship between layers, making it easier to debug. The protocol analyzer's filtering options provide additional debugging capabilities by contracting or expanding the amount and type of information displayed, while maintaining traffic focus.

Signal layer - view any packet on logic analyzer screen.

Event layer - shows the number of occurrences of each event type and the time utilization for each event type, expressed as a total and percent of total

Transaction layer - decodes USB events into transactions.

Data layer - data is collected and can be viewed by pipe (unique address and endpoint). Because the data from each pipe is consolidated, it is also easier to view and analyze. Standard request decoding displays all of the setup / negotiations / capabilities in one easy-to-view location. All data can be viewed in both hex and ASCII.

The USB Protocol Analyzer detects and reports 194 different errors – 11 signaling, 11 packet, 5 framing, and 167 transaction.

Company Name:	Genoa Technology
Address:	5401 Tech Circle
	Moorpark, CA 93021
Telephone:	805-531-9030
Fax:	805-531-9045
Web:	http://www.gentech.com

7.13 Silicon Design Services

Company Name:	MicTron, Inc.
Product Name:	VLSI Design Services

Description: MicTron, Inc. commits to provide world class design services in VLSI technology so that clients can receive quality design works in a timely fashion. They have consulted with a silicon vendor to develop a synthesizable USB device core. Besides the USB device core, they have also applied their expertise in PCI to develop a high performance PCI bus unit for a video conferencing application.

Company Contact:	Mr. Man H. Tam
Company Name:	MicTron, Inc.
Address:	8610 17th Ave. NE
	Seattle, WA 98105
Telephone:	206-545-9449
Fax:	206-545-9449
Email:	mtam@mictron.com

7.14 Simulation Models

Company Name: Phoenix Technologies
Product Name: USB Simulation Test Environment

Features:

USB host simulation model

USB application simulation model

Error insertion and detection features

Transaction-level macro command support

Bus-level macro command support

Simple, well-defined programming interface

Removable module support

Available in Verilog

Maintenance Program support option

Description:

Background

Phoenix Technologies Ltd., a USB industry leader and author of multiple USB device class specifications, presents the Virtual Chips USB Simulation Test Environment. The USB Test Environment (see below) is used to verify compliance with the USB Specification v1.0 and the USB Compliance Checklist. The Test Environment includes a Host Model that supports emulating host USB traffic, and a shell Application Model for simulating FIFOs, sending/receiving data, power management, and querying back-end Function Core signals. The Application Model is designed to be replaced by application logic, as shown below.

Simulation Test Environment Architecture

Chapter 7: USB Development Tools and Helpful Sources

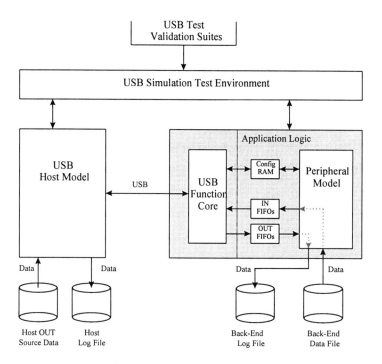

Programming Interface

The Simulation Test Environment includes the Virtual Chips programming interface, which is a set of macro commands to provide the designer with the following levels of control:

Standard-Level Commands – Allows users to generate quickly and easily test variances to determine boundary conditions and diagnose erroneous application logic.

Transaction-Level Commands – Allows users to configure full transaction testing, providing a simple means for designing and executing large, rigorous test scenarios.

Bus-Level Commands – Allows users to model precise bus scenarios specifically focusing on single or multiple error conditions, conducted in serial or parallel, to verify the USB Function.

Host oriented transaction-level macro commands are implemented using bus-level macro commands, which support the following:

Handshake generation

Data IN/OUT generation

Frame generation

Individual J, K line control

Power management control

SYNC bit inversion/truncation error generation

EOP bit inversion/truncation error generation

Bit stuffing error generation

PID mask error generation

CRC-5/CRC-16 error generation

Bit length error generation for PID, data, and CRC

Each bus-level macro command supports standard command parameters such as flags and condition code values, allowing the user to easily configuration bus transactions, inject errors, and determine the status and types of errors returned. In addition, the Host Model also includes numerous utility tasks for error detection, data comparison, and message generation. Thus, the Host Model Programming Interface supports commands that may be used to construct Interrupt Transfers, Bulk Transfers, Isochronous Transfers, Handshake Packets, Power Management Calls, USB Host-Generated Bus Transactions.

Application Model Interface Features

The Application Model is a shell that provides an interface for manipulating FIFOs, sending and receiving data, and handling power management. This shell is designed to be replaced with the Application Logic for a specific peripheral that integrates the USB Function Core.

Phoenix Simulation Test Environment Maintenance Program

Phoenix offers a maintenance program for the USB Test Environment. The maintenance program provides the customer with regular code updates, application notes and errata, and hot line, e-mail, and web-based support.

A Full Range of USB Products

Other USB products include the Virtual Chips USB Open HCI Synthesizable Core, the Virtual Chips USB Hub Synthesizable Core, the Virtual Chips USB Function Synthesizable Core, and Phoenix BIOS USB System BIOS Extensions.

Company Name: Phoenix Technologies, Ltd.
Address: 411 E. Plumeria
San Jose, CA 95134
Telephone: 408-570-1000
Fax: 408-570-1001
Web: www.phoenix.com

Company Name: Sand Microelectronics, Inc.
Product Name: USB Simulation Model
Features:

Comprehensive model of the Universal Serial Bus Revision 1.0

Facilitates the verification of USB host, hub, and device designs

Verilog/VHDL source code of the model provided for maximum customer flexibility

Model controlled through user-written Verilog/VHDL code

Three independently controllable models:

    Host

    Device

    Monitor

Simulation Models

Digital PLL integrated into the host, device, and monitor models

Enhanced monitor

    Logs transactions on the USB

    Protocol checks

Support option available

Optional USB compliance test suite

Description: Sand's USB Simulation Model is a behavioral model that simulates the transactions of the Universal Serial Bus. The model, which is provided in Verilog/VHDL source code format, is a tool for system designers to exercise and debug the design of components and systems based on USB. The objective of the model is to aid in the functional verification process and to reduce functional and/or timing errors prior to silicon or board fabrication.

The model is powerful, flexible, and easy to use. A procedural interface is provided whereby users can control the model through user-written Verilog/VHDL code.

The model checks for protocol of all USB transactions and flags any errors. In addition, special logging features are incorporated into the model to facilitate the debug process.

This product is a comprehensive model that supports all transactions and transfers in the USB Revision 1.0 specification. In addition, the model provides a means of simulating various error conditions on the USB both in terms of transmission as well as protocol.

The model is comprised of three individually controllable sub-models (Host, Device, and Monitor), each of which is described below.

USB Host

The USB Host model simulates transactions initiated by a USB Host Controller and can be used to verify the functionality of a USB Device or USB Hub design. The following USB transactions are supported:

Bulk

Isochronous

Interrupts

Control

Reset

Resume

Time Stamp

The USB Host model provides two basic command types to the user: Transaction and Modify commands. Transaction commands generate one or more USB transfers on the bus. At the completion of the transaction the model returns status of the transaction which can be used as a powerful steering mechanism in creating dynamic test sequences. The Modify commands enable the user to dynamically change the attributes pertaining to the host model during run time. These commands do not consume any simulation time The USB Host model provides support for high/low speed devices as well as USB Hubs.

USB Device

The USB Device model simulates a fully-functional USB Device and responds to transactions initiated by a USB Host. It can be used to verify the functionality of a USB Host or USB Hub design. Users can modify USB Device behavior by programming the descriptors such as device, configuration, interface, and endpoint. The features supported by the USB Device are:

Low-speed or high-speed operation

Supports 16 endpoints in high-speed

Supports multiple configurations and interfaces

Error Generation/Checking

The USB Host and USB Device models provide a mechanism for generating and/or checking error conditions on the bus as described below:

CRC errors

Bit stuff errors

Synthesizable Cores

Sync field errors

EOP errors

Token errors

Data/Token PID errors

Data toggle errors

Handshake errors

Byte boundary errors

Error and transaction logging

Clock errors

USB Monitor

The Monitor model provides the monitoring of all transactions on the USB, performing logging and protocol monitoring. It tracks all the transmission and packet level errors and provides for protocol error checking.

Company Name:	Sand Microelectronics, Inc.
	3350 Scott Boulevard, #24
	Santa Clara, CA 95054
Telephone:	(408) 235-8600
Fax:	(408) 235-8601
Email:	sales@sandmicro.com
Web:	http://www.sandmicro.com

7.15 Synthesizable Cores

Company Name:	Decicon Incorporated
Product Name:	USB Synthesizable Function Core

Features:

Modular, reconfigurable architecture

- Up to 15 input and 15 output endpoints plus endpoint 0
- Dynamically configurable endpoints
- Scalable number of interfaces
- Scalable number of configurations
- Low- and high-speed versions available
- Simple interface for endpoint FIFOs
- USB Standard Command Processor
- USB class/vendor specific command interface
- Power management functions supported

Description: USB Synthesizable Function Core is a collection of modular function blocks. This modular architecture allows the user to configure the USB Core according to system requirements. In cases where the application has the capability to support the USB functions, the user can remove some of the blocks in order to minimize the gate count of the USB Core. These blocks include the USB Standard Command Processor and the Descriptor ROM Control Block.

In the most comprehensive implementation of the core, all the USB Standard Commands can be handled internally and a simple interface is provided for the transfer of the class/vendor specific commands to the application logic. In implementations where the USB Standard Command Processor has been removed, this interface block also handles the transfer of standard device commands to the application logic.

The design can be easily modified to have the required number and type of endpoints, with very little or no redundant logic. All operational parameters of the endpoints can be reconfigured during the operation of the block, allowing complex data transfer modes.

The descriptor data is stored in the External Descriptor ROM, and the flexible structure of the ROM Control Block allows the user to implement the core with any number of configurations and interfaces. Class/vendor specific descriptors can also be stored in the ROM, along with the standard descriptor data.

Synthesizable Cores

USB Core requires a single clock input with a frequency of 48 MHz for fast and 6 MHz for slow devices. This clock is used directly by the Digital PLL. The clock for the rest of the core is extracted by the PLL from the USB, and has a frequency of 12 MHz for fast and 1.5 MHz for slow devices.

USB Core supports SUSPEND/RESUME signaling. The REMOTE_WAKEUP feature is also supported.

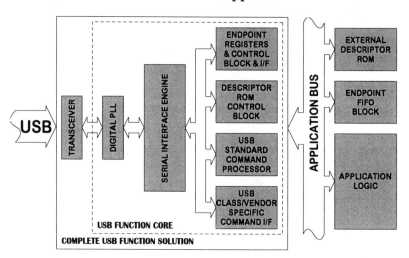

Company Name: Decicon Incorporated
Address: 1250 Oakmead Parkway, Suite 316
Sunnyvale, CA 94086, USA
Telephone: (408) 720-7690
Fax: (408) 720-7691
Email: info@decicon.com
Web: www.decicon.com

Company Name: Oki Semiconductor
Product Name: USB Hub Mega Macrofunction

Model Number: W722

Features:

USB v1.0 compliant

Supports up to four downstream ports (expandable)

Can replace external downstream ports with internal ports to interface with embedded functions

Supports any combination of low-speed (1.5 Mb/sec) and full-speed (12 Mb/sec) downstream ports

Supports one control endpoint and one interrupt endpoint

Auto sensing capability for power sources: can behave as either self-powered or bus-powered hub

Individual port power switching, overcurrent protection, and overcurrent indicator

Supports remote-wakeup by the Hub itself, the embedded function, or downstream devices

Description: Oki's Universal Serial Bus (USB) Hub Mega Macrofunction (W722) is a featured element in Oki's 0.5µm Sea of Gates (SOG) and Customer-Structured Array (CSA) families.

The W722 provides Oki's Serial Interface Engine (SIE), a Hub Core Controller (HCC), a Hub Repeater (HR), status/descriptor/register file block, and an optional FIFO controller/application interface for an embedded function in five highly integrated submodules. The submodule partitioning allows custom configurations to be easily developed.

Synthesizable Cores

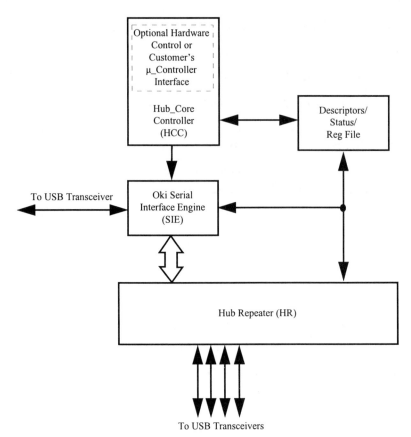

W722 Block Diagram (stand alone)

Oki's Serial Interface Engine (SIE) handles the USB communication protocol. It performs clock generation, packet sequencing, signal generation/detection, CRC generation/checking, bit-stuffing and PacketID generation/decoding. When the W722 is used to implement a compound device, the Oki Serial Interface Engine (SIE) can be shared by both the Hub function and the embedded function.

Chapter 7: USB Development Tools and Helpful Sources

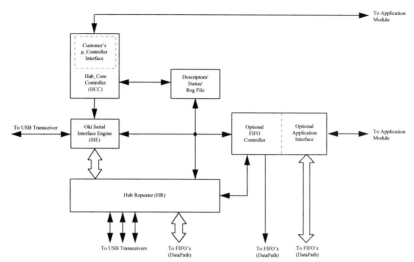

W722 Block Diagram (compound device)

The Hub Core Controller (HCC) includes:

a. A Request Parser to interpret the Host Requests/Tokens to both the default Endpoint 0 and the Status Change Endpoint 1.

b. A DMA controller/pointer block to handle data movement from/to the status descriptor memory and the register file. It is capable of handling aborts and retries.

c. A parallel read/write interface to a customer's micro-controller.

d. Optional control State machines to execute requests, respond to tokens, and handle errors if customers choose not to use a micro-controller.

The Hub Repeater (HR) includes:

a. Repeater logic to manage connectivity on a per packet basis. It can handle any combination of full-speed and low-speed devices at the downstream ports. It also supports exception handling such as fault recovery, suspend/resume as directed by the Host, as well as remote-wakeup and Frame timer synchronization.

Synthesizable Cores

  b. Four Port State Machines, which can interpret and respond to bus events such as connect/disconnect detection, port enable/disable, suspend/resume, reset, and power switching.

  c. The power manager and central clocking circuitry.

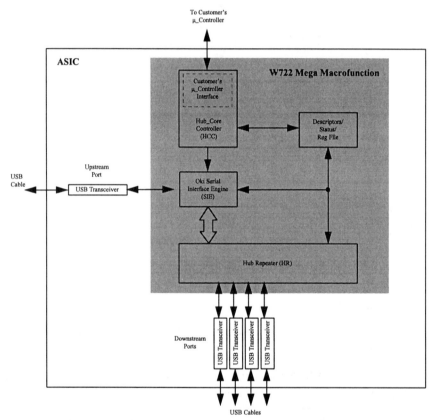

Example USB Hub Macrofunction Application (stand alone)

The Port State Machine logic can be implemented as an external port, or an internal port for communication with an embedded function. To expand the Hub Mega Macrofunction, additional Port State Machine functional blocks can be added.

Chapter 7: USB Development Tools and Helpful Sources

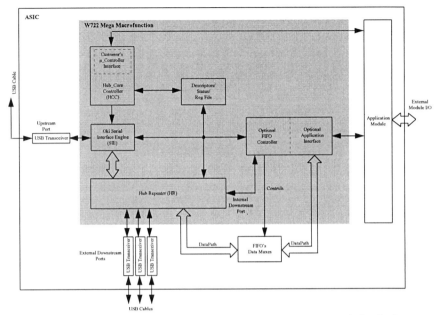

Example USB Hub Macrofunction Application (compound device)

The Descriptors/Status/Register File block includes standard and Hub descriptors, Hub and Port status, and the Register file to store all control and status information.

The FIFO controller/application interface block manages all FIFO operations for the embedded functions. It provides a parallel read/write interface to the customer's application. It supports up to eight asynchronous FIFOs, four transmit and four receive. They can be configured as described in the table below.

FIFO Configuration

FIFO Type	Endpoint Address	Programmable	Function
Transmit	0	64 bytes	Control Transfers
Transmit	5	64 bytes	Interrupt and Bulk Transfers
Transmit	6	64 bytes	Interrupt and Bulk Transfers
Transmit	7	2 Kbytes	Isochronous, Interrupt, and Bulk Transfers
Receive	0	64 bytes	Control Transfers
Receive	1	64 bytes	Bulk Transfers
Receive	2	64 bytes	Bulk Transfers
Receive	3	2 Kbytes	Isochronous and Bulk Transfers

Synthesizable Cores

Endpoint 3 and 7 are two-level FIFOs, which support up to two separate data sets of variable sizes. All FIFOs have flags that detect full and empty conditions and have the capability of re-transmitting or re-receiving the current data set. The FIFO controller logic can be customized to support various FIFO/endpoint requirements.

The W722 connects to the Universal Serial Bus via Oki's universal USB transceiver. The USB specific I/O converts the W722's internal unidirectional signals into USB compliant signals. The universal USB transceiver allows the designers' application module to interface with the physical layer of the Universal Serial Bus. It transmits and receives serial data at both full-speed (12Mb/s) and low-speed (1.5Mb/s) data rates.

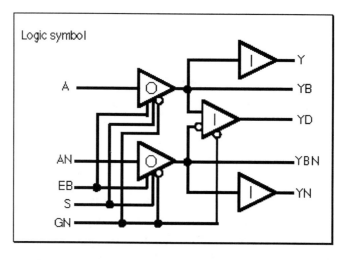

BUD2SLL: USB I/O Buffer with Full/Low Speed

Company Name: Oki Semiconductor

Product Name: USB Device Controller Macrofunction

Model Number: W712

Features:

Chapter 7: USB Development Tools and Helpful Sources

USB v1.0 compliant

Full-speed (12 Mb/sec) and low-speed (1.5 Mb/sec) support

Parallel read/write application interface

Supports isochronous, control, interrupt and bulk transfers

Supports four transmit FIFO's

    Three 64 byte

    One 2 Kbyte (2-level)

Supports four receive FIFO's

    Three 64 byte

    One 2 Kbyte (2-level)

Supports one control endpoint and six additional endpoint addresses

Expandable up to 32 endpoint addresses

Customizable to specific application requirements

Description: The Universal Serial Bus (USB) Device Controller Mega Macrofunction is a featured element in Oki's 0.5µm Sea of Gates (SOG) and Customer Structured Array (CSA) families.
Oki's USB Mega Macrofunction provides a USB interface, control/status block, FIFO control, and application interface in two highly integrated submodules for system design interfaces based on the USB protocol. The submodule partitioning allows custom configurations to be easily developed.

Synthesizable Cores

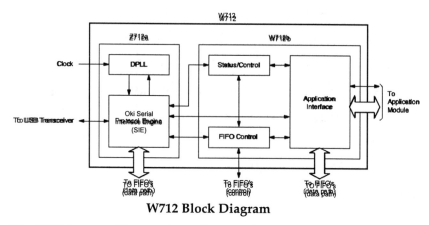

W712 Block Diagram

The W712 controller consists of two submodules, the Z712a hard macro, and the W712b soft macro, each containing multiple function blocks. The Z712a includes Oki's Serial Interface Engine (SIE), DPLL, and Timer Blocks. The W712b includes the Status/Control, FIFO Control, Application Interface, Frame Timer Synthesizer, and remote wakeup blocks.

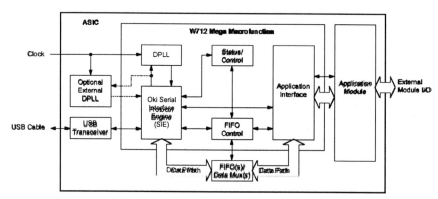

Example USB Device Macrofunction Application

The USB mega macrofunction connects an industry standard USB interface with a parallel read/write application interface. This straightforward interface permits easy integration of the USB mega macrofunction to the target application. Using Oki's USB mega macrofunction, designers can reduce development time, risk, and introduce their USB based products to market faster. Oki's W712

Chapter 7: USB Development Tools and Helpful Sources

USB Device Controller mega macrofunction provides a complete USB device interface solution and is fully compliant with the Universal Serial Bus 1.0 specification.

Oki's Serial Interface Engine (SIE) handles the USB communication protocol. It performs packet sequencing, signal generation/detection, CRC generation/checking, NRZI data encoding, bit stuffing and packet ID (PID) generation/decoding.

The Digital Phase Locked Loop extracts the clock and data from the USB differential received data.

The Timer block monitors idle time on the USB bus.

The Status/Control block uses transfer type and FIFO state information to manage the reception and transmission of USB data. It monitors the transaction status and communicates control events to the application via the Application Interface.

The FIFO control block manages all FIFO operations for transmitting and receiving USB data sets. The W712 supports eight FIFOs (four transmit and four receive). They can be configured as described in the table below.

FIFO Configuration

FIFO Type	Endpoint Address	Programmable	Function
Transmit	0	64 bytes	Control Transfers
Transmit	5	64 bytes	Interrupt and Bulk Transfers
Transmit	6	64 bytes	Interrupt and Bulk Transfers
Transmit	7	2 Kbytes	Isochronous, Interrupt, and Bulk Transfers
Receive	0	64 bytes	Control Transfers
Receive	1	64 bytes	Bulk Transfers
Receive	2	64 bytes	Bulk Transfers
Receive	3	2 Kbytes	Isochronous and Bulk Transfers

Endpoint 3 and 7 are two-level FIFOs, which support up to two separate data sets of variable sizes. All FIFOs have flags that detect full and empty conditions and have the capability of re-transmitting

Synthesizable Cores

or re-receiving the current data set. The FIFO controller logic can be customized to support various FIFO/endpoint requirements.

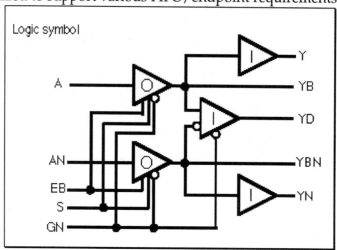

BUD2SLL: USB I/O Buffer with Full/Low Speed

The W712 connects to the Universal Serial Bus via Oki's universal USB transceiver. The USB specific I/O converts the W712's internal unidirectional signals into USB compliant signals. The universal USB transceiver allows the designers' application module to interface with the physical layer of the Universal Serial Bus. It transmits and receives serial data at both full-speed (12 Mb/s) and low-speed (1.5 Mb/s) data rates.

Company Name: Oki Semiconductor
Address: 785 North Mary Ave.
Sunnyvale, CA 94086
Telephone: 408-720-1900
Fax: 408-720-1918
Email: lau@okisemi.com
Web: www.okisemi.com

Company Name: Phoenix Technologies, Ltd.
Product Name: USB Function Synthesizable Core

Chapter 7: USB Development Tools and Helpful Sources

Features:

Device support:

- Human interface device support
- Printer support
- Motion imaging class support
- Still-imaging class support
- Communication class support
- Mass storage class support

Other features:

- Optimized serial interface engine
- Phoenix BIOS compatible
- Supported by Virtual Chips USB Simulation Test Environment
- Maintenance program support option
- Available in Verilog

Description:

Background

Phoenix Technologies Ltd., a USB industry leader and author of multiple USB device-class specifications, presents the Virtual Chips USB Function Synthesizable Core. This core is USB v1.0 compliant, and is designed to accommodate devices that range in complexity from low-speed human input devices to high-speed communication and imaging devices. The core provides the following benefits:

- Most logic on a single clock domain
- Scalable number of endpoints, number of interfaces, packet size, and ROM size
- Removable module support
- Dynamically re-configurable endpoints
- Microcontroller application interface support

Power management support

Asynchronous and synchronous FIFO support (multiple back-end endpoint data buffer interface support)

Digital phase lock loop clock recovery

External configuration ROM support

USB standard command processor

USB class-specific and vendor-specific command interface

USB serial interface engine

Scalable Endpoint Support

The USB Function Core provides a programmable feature to configure 1, 2, 4, 8 or 16 endpoints, allowing customers to tailor the core to their own design needs. By removing artificial restrictions on core flexibility Phoenix allows customers to minimize gate count and differentiate their peripherals based on individual design and product requirements.

Scalable Interface Support

The USB Function Core provides a programmable feature to configure 1, 2, 4, 8, or 16 interfaces. This feature offers a variable number of USB Interface Descriptors and Endpoint Descriptors, and even allows endpoints to be reprogrammed dynamically to accommodate complex devices with multiple data transfer modes.

Removable Module Support

The USB Function Core allows the designer to implement configuration and command support in firmware or application logic. To implement support in firmware, the Configuration Scan Block, Standard Command Decoder Block, and Endpoint Register Block may be removed.

Application Interface Support

The USB Function Core application interface includes I/O signals for FIFO, class and vendor command parsing, and error condition handling. The Function Core may be used with standard

Chapter 7: USB Development Tools and Helpful Sources

microcontrollers, DSP controllers, and device-specific state-machines (see reverse for the Function Core Architecture).

Auto-Configuration Support

The USB Function Core provides auto-configuration services to manage USB device requests sent to endpoint 0, eliminating extensive microcontroller code. The core automatically processes standard device requests, including:

CLEAR_FEATURE	SET_ADDRESS
GET_CONFIGURATION	SET_CONFIGURATION
GET_DESCRIPTOR	SET_FEATURE
GET_INTERFACE	SET_INTERFACE

Additional I/O signals provide class- and vendor-specific command support, allowing application logic to directly process USB device requests.

Phoenix Core Maintenance Program

Phoenix offers a maintenance program for the USB Function Core. The maintenance program provides a customer with regular core updates for compatibility and errata, for application notes, and hot line, e-mail, and web-based support.

A Full Range of USB Products

Other Virtual Chips USB products include the Virtual Chips USB Open HCI Synthesizable Core, the Virtual Chips USB Hub Synthesizable Core, the Virtual Chips USB Simulation Test Environment, and Phoenix BIOS USB System BIOS Extensions.

Company Name: Phoenix Technologies, Ltd.
Address: 411 E. Plumeria
San Jose, CA 95134
Telephone: 408-570-1000
Fax: 408-570-1001
Web: www.phoenix.com

Company Name: Phoenix Technologies, Ltd.

Product Name: USB OHCI Host Synthesizable Core

Features:

USB v1.0 Compliant

Open HCI v1.0 Compliant

33MHz PCI v2.1 Compliant

Windows 95 Compatible

Technology-independent design

Silicon-proven

Integrated root hub

12 Mb/s data transfer rate

1.5 Mb/s data transfer rate

Legacy keyboard/mouse support

Supports 127 devices

Supports Control, Bulk, Isochronous, and Interrupt data transfer types

Approx. 30K gates

Maintenance Program support option

Description:

Background

Phoenix Technologies Ltd., a USB industry leader and author of multiple USB device class specifications, presents the Virtual Chips USB Open Host Controller Interface Core. The core is a complete PCI to USB interface and was designed and verified by Compaq Computer. The core is fully compliant with the USB Specification and the USB Open HCI Specification authored by Microsoft, Compaq, and National Semiconductor. The core may easily be used for:

Motherboard, PCI-based chipset designs

Stand-alone PCI-based standard ASSP designs

The design, shown below, includes a 33 MHz PCI interface, and a Host Controller and a USB Interface.

USB Interface

The USB Interface contains two primary modules: the Serial Interface Engine (SIE), responsible for the bus protocol, and the Root Hub, used to expand the number of USB ports. The SIE performs the following functions:

CRC coding	Parallel to serial conversion
Serial to parallel data conversion	NRZI encoding
Bit stuffing	SOF token generation

Synthesizable Cores

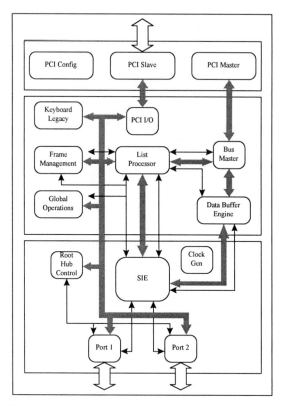

Host Controller

The USB Host Controller incorporates much of the intelligence required for processing incoming and outgoing data, as well as legacy keyboard support for PS/2 keyboards connected to PS/2 keyboard controllers via Port 60/64. The USB Host Controller includes logic for:

Frame management

List processing

Power management

Packet validation

Phoenix OHCI Core Maintenance Program

Phoenix offers a maintenance program for the USB Host Synthesizable Core. The maintenance program provides a customer with regular core updates for compatibility and errata, for application notes, and hot line, e-mail, and web-based support.

A Full Range of USB Products

Other USB products include the Virtual Chips USB Function Synthesizable Core, the Virtual Chips USB Hub Synthesizable Core, the Virtual Chips USB Simulation Test Environment, Phoenix MultiKey Keyboard Controller BIOS, and Phoenix BIOS USB System BIOS Extensions.

Company Name:	Phoenix Technologies, Ltd.
Address:	411 E. Plumeria
	San Jose, CA 95134
Telephone:	408-570-1000
Fax:	408-570-1001
Web:	www.phoenix.com

Company Name: Sand Microelectronics, Inc.

Product Name: USB Device Controller Synthesizable Core

Features:

Silicon proven, USB device synthesizable core, USB 1.0 compliant

Has been compatibility tested at the USB Plugfests

Supports wide range of applications: pointing devices, scanner, camera, fax, printer, speaker, modem, monitor, etc.

Verilog/VHDL source code provided

Supports low speed and full speed devices

Provides simple read/write parallel interface to application

Integrated digital PLL

Programmable number of endpoints

Easily configurable endpoint organization using Rapidscript configuration utility

Maintains address pointer for Endpoint 0 transactions

Supports up to three configurations, up to four interfaces per configuration, up to four alternate settings per interface

Supports all USB standard commands

Easy-to add vendor/class commands

Suspend/Resume logic provided

Approximately 6k gates (five endpoints)

Description: Sand Microelectronics' USB Device Controller Synthesizable Core (UDC Core) is a set of synthesizable building blocks that ASIC/FPGA designers can use to implement a complete USB Device interface. The UDC Core is fully compliant with Revision 1.0 of the USB specification. By utilizing Sand's silicon proven UDC Core, designers can significantly reduce development time and engineering risk and bring their USB based solutions to market faster. The UDC Core contains all the necessary hooks for customizing and optimizing it for a specific application. In addition, the design can be easily migrated to almost any technology in a relatively short period of time.

Sand's UDC Core bridges an industry standard USB interface to a simple read/write parallel interface. The simple interface shields the designer from the complexities of the USB protocol and makes it easy to integrate Sand's UDC Core to the Customer's target application. The UDC Core runs on a 12 MHz clock, which is extracted from the 48 MHz clock provided by the user. The UDC Core is comprised of four main blocks as described below:

Phase Lock Loop (PLL) Block: The PLL Block is a Digital Phase Lock Loop, which extracts the Clock and Data from the USB Cable. The input to the PLL Block is from the USB Differential Transceiver. The PLL runs on a 48 MHz clock. The PLL Block also generates a 12 MHz clock from the 48 MHz clock which it supplies to the SIE and UBL Blocks. The PLL identifies the SE0 signal on the USB and sends it to the SIE Block.

Serial Interface Engine (SIE) Block: The SIE block does all the front end functions of the USB protocol such as SyncField Identification, NRZI-NRZ Conversion, Token Packet decoding, Bit Stripping, Bit Stuffing NRZ-NRZI conversion, CRC5 checking and CRC16 generation and checking. The SIE Block also converts the serial data coming in the Data Packet into 8 bit parallel data. The SIE Block has a one deep 8 bit FIFO built-in for buffering the data during data transmission and reception.

USB Bridge Layer (UBL) Block: The UBL block handles the error recovery mechanism during transactions and interfaces to the Device Logic. The UBL also handles all the Standard Control Transfers addressed to Endpoint 0. The UBL Block is further sub-divided into the Protocol Layer (PL) and Endpoint (EP) blocks. The PL Block controls the SIE Block by providing necessary handshake signals, and communicates with the Device Interface Logic. It also has the mechanism for error recovery if the Device Interface violates the data transfer protocol. The EP Block handles all the USB standard command, and passes the USB class and vendor commands to the application. The UBL supports up to three configurations, with each configuration having a maximum of four interfaces. Each interface can have up to four alternate settings.

EPINFO Block: This block can be configured by the user for various applications using Sand's Rapidscript configuration utility. Information related to endpoints, configuration, and string descriptors are stored in this block. EPINFO Block contains the DATA0/DATA1 synchronization bits for each endpoint and supports up to 16 active bi-directional logical endpoints. The number of physical endpoints is programmable.

Company Name: Sand Microelectronics, Inc.

3350 Scott Boulevard, #24

Santa Clara, CA 95054

Telephone: (408) 235-8600, Fax: (408) 235-8601

Email: sales@sandmicro.com

Web: http://www.sandmicro.com

Company Name: Sand Microelectronics, Inc.

Product Name: USB Host Controller Synthesizable Core

Features:

Silicon proven, USB host controller core

USB 1.0 compliant

Compatible with open HCI 1.0 specification and Windows'95 USBD

Verilog/VHDL source code provided

Supports low-speed and full-speed devices

Configurable root hub

    Up to fifteen downstream ports

Simple application interface facilitates:

    Bridging core to other system buses such as PCI

    Integration of the core with chipsets and microcontrollers

Integrated digital PLL

Supports for SMI interrupts

Approximately 22K gates

Description: Sand Microelectronics' USB Host Controller Synthesizable Core (UHOSTC Core) is a set of synthesizable building blocks that ASIC/FPGA designers can use to implement a complete USB Host Controller function. The UHOSTC Core is fully compliant with USB 1.0 and Open HCI 1.0 specifications. By utilizing Sand's silicon proven UHOSTC Core, designers can significantly reduce development time and engineering risk and bring their USB based solutions to market faster. The UHOSTC Core contains all the necessary hooks for customizing and optimizing it for a specific application. In addition, the design can be easily migrated to almost any technology in a relatively short period of time.

Sand's UHOSTC Core can be easily bridged to industry standard buses such as PCI. A simple application interface is provided which shields the designer from the complexities of the USB Host controller and makes it easy to integrate Sand's UHOSTC Core to the Customer's target application.

The UHOSTC Core is comprised of the following main blocks as described below:

HCI Slave Block

This block contains the OHCI operational registers, which are programmed by the Host Controller Driver (HCD).

HCI Master Block

The HCI Master Block handles all the reads/writes to system memory that are initiated by the List Processor while the Host Controller (HC) is in the Operational State and is processing the lists queued in by HCD. It generates the addresses for all the memory accesses. The major tasks handled by this block are:

Fetching endpoint descriptors (ED) and transfer descriptors (TD)

Read/Write endpoint data from/to system memory

Accessing HC communication area (HCCA)

Write status and retire TDs

USB State Control

This block implements the USB Operational States of the Host Controller as defined in the OHCI Specification. It also generates SOF tokens every millisecond and triggers the List Processor while HC is in the Operational States.

Data FIFO

This block contains a 64x8 FIFO to store the data returned by Endpoints on IN tokens and the data to be sent to the Endpoints on OUT Tokens. The FIFO is used as a buffer in case the HC doesn't get timely access to the host bus.

List Processor Block

Synthesizable Cores

The List Processor processes the lists scheduled by HCD according to the priority set in the operational registers.

Root Hub & Host SIE

The Root Hub propagates Reset and Resume to down-stream ports and handles port connect/disconnect. The Host Serial Interface Engine (HSIE) converts parallel to serial, serial to parallel, NRZI encoding/decoding, and manages USB serial protocol.

Company Name: Sand Microelectronics, Inc.

3350 Scott Boulevard, #24

Santa Clara, CA 95054

Telephone: (408) 235-8600

Fax: (408) 235-8601

Email: sales@sandmicro.com

Web: http://www.sandmicro.com

Company Name: Sand Microelectronics, Inc.

Product Name: USB Hub Synthesizable Core

Model Number: UH01

Features:

Silicon proven, USB hub synthesizable core, USB 1.0 compliant

Has been compatibility tested at the USB Plugfests

State machine design - no microcontroller or firmware required

Verilog/VHDL source code provided

Supports low- speed and full-speed devices on downstream ports

Integrated digital PLL for clock and data recovery from USB

User-configurable options via hub Rapidscript:

- Downstream ports from two to 15

Chapter 7: USB Development Tools and Helpful Sources

- Port power switching mode and overcurrent protection

- Number of string descriptors

Downstream device connect/disconnect detection

Supports Suspend/Resume for Power Management

Supports one interrupt endpoint in addition to Endpoint 0

Approximately 12K gates for four downstream port implementation

Description: Sand Microelectronics' USB Hub Synthesizable Core (UH01 Core) is a set of synthesizable building blocks that ASIC/FPGA designers can use to implement a complete USB Hub. The UH01 Core is fully compliant with Revision 1.0 of the USB specification. By utilizing Sand's silicon proven UH01 Core, you can significantly reduce development time and engineering risk and bring your USB based hub solutions to market faster. The UH01 Rapidscript Utility enables designers to easily configure the core (such as setting the number of down stream ports, for example). In addition, the core design is provided in Verilog/VHDL making it easy to synthesize into any ASIC/FPGA technology.

Sand's UH01 Core consists of the Hub Repeater and the Hub Controller. The Hub Repeater is responsible for connectivity setup and tear-down and supports exception handling such as bus fault detection/recovery and connect/disconnect detect. The Hub Controller provides the mechanism for host to hub communication. Hub specific status and control commands permit the host to configure a hub and to monitor and control its individual downstream ports. The UH01 Core is comprised of following main blocks as described below:

Hub Controller Block (HUBCTRL): This block contains the Digital PLL (to extract the clock and data from the USB cable), Serial Interface Engine (to handle the bit level protocol on the USB, to convert serial data to parallel and vice-versa), and a Command Interpreter (HUBCI). The HUBCI decodes the USB standard commands, hub class commands and provides the necessary Hub/Port status information to the host. In addition, the HUBCI contains the USB standard descriptors such as device descriptor, configuration descriptor, interface descriptor, endpoint descriptors and hub descriptors. The descriptor data is provided to the host when requested. The HUBCI maintains the HubStatus and HubChange registers, the PortStatus and PortChange information, and the status change bit map for the interrupt endpoint to report any change in the hub/port status.

Hub Repeater Block (HUBRPT): This block handles the connectivity between the root port and downstream ports. It takes the D+ and D- signals from the root port and all the downstream ports and establishes the connectivity between one downstream port and the root port or from root port to all the downstream ports based on current state of the port state machine. A multiplexer is used to select the data coming from the downstream ports for upstream connectivity.

Frame Timer Block: This block contains the USB Hub Frame Timer logic. It generates the EOF1 and EOF2 points in all frames, which are used by the Hub Repeater Block for establishing the connectivity between the ports.

Port State Machine Block (PRTSM): This block controls the downstream port. There is one block per downstream port. The PRTSM detects the connect/disconnect events on the port and has the capability to enable/disable/suspend the port. The PRTSM reports the port status and change information to the HUBCI Block for the corresponding port with which this PRTSM Block is associated. The PRTSM detects whether the device attached is a full-speed or low-speed device. Based on the type of device attached, this block converts the full speed signaling from the root port to full-speed/low speed signaling to the downstream port and vice-versa. This block also generates the low-speed keep-alive strobes if a low-speed device is connected to the port.

Power Switching and Overcurrent Control Block (PWRSWTCH): This block contains the power switching mode and the overcurrent protection control logic. This block is programmable by the user to support different power switching modes and overcurrent protection modes. This block controls the power switching for all the Port State Machines and monitors the overcurrent condition from all downstream ports.

Suspend Resume Block: This block contains the hub functional state machine and handles all the USB suspend/resume events both as a hub as well as a USB device.

The USB Hub Core generates the necessary signals to interface to any standard USB transceiver part.

Company Name: Sand Microelectronics, Inc.
3350 Scott Boulevard, #24
Santa Clara, CA 95054
Telephone: (408) 235-8600, Fax: (408) 235-8601
Email: sales@sandmicro.com
Web: http://www.sandmicro.com

7.16 Transceiver Design

Company Name: Decicon Inc.
Product Name: USB Transceiver for Embedded Applications
Features:

USB Specification Rev. 1.0 compliant

Supports low- and high-speed transmissions

Can be targeted to virtually any CMOS process, to be integrated along with the rest of the USB functionality

Description: The USB Transceiver is developed to act as the interface between the USB digital core and the differential USB. The present design is in full compliance with the Universal Serial Bus Specification, Revision 1.0 and supports both full-speed and low-speed transmissions. The design can be targeted to virtually any CMOS process. The transceiver is available in the forms of electrical design or physical design in customer's process of choice.

Company Name: Decicon Incorporated

Address:	1250 Oakmead Parkway, Suite 316
	Sunnyvale, CA 94086, USA
Telephone:	(408) 720-7690
Fax:	(408) 720-7691
Email:	info@decicon.com
Web:	www.decicon.com

7.17 Workshops, Conferences, and Books

Company Name: Annabooks

Description: Annabooks provides USB designers with a variety of books, workshops, and developer's conferences. Books cover subjects ranging from architectural hardware and software issues to specific design information for engineers who need to comply with the specification without necessarily being concerned with the underlying architectural issues. Workshops are scheduled periodically in various locations so engineers can learn from experienced designers and architects. Typical workshops take two days. Conferences take place in the US in the spring and Europe in the fall. Plans are also being made for at least one conference in Asia each year. Up-to-date schedules of events and latest listings of books are available on Annabooks' web site.

Company Name:	Annabooks
	11838 Bernardo Plaza Ct.
	San Diego, CA 92128
Telephone:	800-462-1042
	619-673-0870
Fax:	619-673-1432
Email:	info@annabooks.com
Web:	http://www.annabooks.com

8. Appendix

8.1 USB Power Application Notes

This section is from a technical paper by Jonathan M. Bearfield, Power Supply Components Specialist at Texas Instruments. Permission to reprint this material is gratefully acknowledged. Questions to Mr. Bearfield may be directed to him through info@annabooks.com.

8.1.1 Purpose of this section

This paper provides an introduction to the Universal Serial Bus (USB) interface as it relates to the power distribution requirements of the voltage bus.

The general requirements for USB power distribution set up the need for an in-line device on the voltage bus that can either switch the voltage to the bus or provide overcurrent protection for the voltage bus. Implementation of this allows for several device configurations that include, but are not limited to, MOSFET switches, PTC resistors (polyfuses), one time fuses, and/or solid state switches.

The basic requirements of the USB specification do not guarantee subsequent operation of any port once the overcurrent protection device recognizes a fault condition. After rectifying the fault condition in the system, some re-initialization and/or maintenance may be required in order to bring the system back up to a fully operational state.

There are many requirements in the current USB specification (Version 1.0) concerning voltage regulation and the current limits of the system. This paper discusses the issues of the specification

233

concerning real world limitations and requirements for the voltage and current requirements of the USB voltage bus.

8.1.2 Introduction

The USB interface is a 12 Mb/s multiplexed serial bus designed for low- to medium-speed PC peripherals. USB utilizes asynchronous and isochronous data transmission. USB is a four-wire interface conceived for dynamic attach-detach (hot plug-unplug) of peripheral devices in the PC environment.

The power distribution section of the USB specification breaks down into five separate areas for discussion: device class requirements, overcurrent protection, connectivity limitations, power supply voltages, and voltage regulation requirements. Each of these areas interrelates with the others, and changes or considerations for any one of the areas will affect the limitations imposed by the others. When evaluating the specification for power requirements, read Sections 6 and 7 of the current USB specification, at a minimum.

The major concerns when designing a USB hub and selecting a power distribution device are the allowable voltage drop across the overcurrent protection device, and the trip current threshold. Some of the limits posed on the voltage bus actually come from other governing specifications like the UL specification.

In addition to meeting all the required specifications, the interface must be a practical design. The trade off made when designing the USB power distribution interface involves the difficulties that arise when trying to design in practical end user features or responses; features which make the system a practical interface as well as a functional interface over the life of the end product.

Some practical additional features from the end user's point of view may include thermal protection, switched power out of the high-powered ports, visual-software-initiated responses, and maintenance free overcurrent protection for faults on the voltage bus. Finding a cost-effective solution that meets the needs of both

Chapter 8: Appendix

the USB specification and the user interface is where that design challenge resides.

8.1.3 Classes Of Devices

The USB specification provides for five basic device classes. These are: Bus-Powered Hub, Self-Powered Hub, Low-Power/Bus-Powered Function, High-Power/Bus-Powered Function, and Self-Powered Function. For the purpose of this discussion we will elaborate on each of the classes solely based on the voltage and/or current that system and/or function can consume and/or distribute.

Bus-Powered Hub

This hub draws all of the power for all internal functions and output ports, for downstream loads, from its USB voltage bus input. A maximum of 500 mA may be drawn by a bus-powered hub. It will supply 100 mA (max.) to any downstream ports, and may consume any portion of the 500 mA, but is limited to draw no more than 100 mA at power-up.

Self-Powered Hub

In this case, the power for the internal functions and downstream ports does not come from the USB voltage bus. The self-powered hub must be capable of supplying 500mA to each of the downstream ports that it maintains. It may draw 100 mA from the bus to provide power to the USB interface. A host, by definition, is a self-powered hub.

Low Power, Bus-Powered Function

All power to this device comes from the USB voltage bus. It may draw a maximum of 100mA (max.) during normal operation.

High Power, Bus-Powered Function

All power to this device comes from the USB voltage bus. It may draw up to 100 mA (max.) during power up and may draw up to 500 mA (max.) during normal operation.

USB Power Application Notes

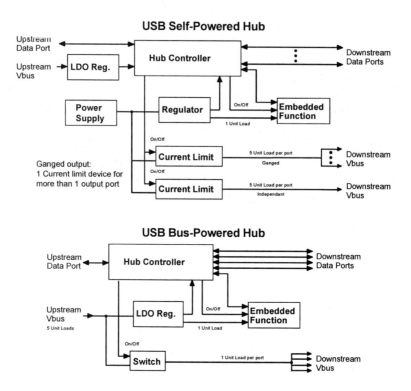

Each of the classes mentioned has specific current sourcing/sinking requirements and voltage regulation requirements. Every USB device must meet the regulation and voltage bus requirements regardless of the device class.

8.1.4 Protection Requirements

Self-Powered Hub Power Distribution Design

Host and SP (self-powered) hub devices receive power from and internal power supply. They each must implement overcurrent protection on each of the downstream port connections for safety reasons. No single port may deliver more than 5.0 Amps to the voltage bus or into a fault and still meet the current UL regulatory safety limits. Therefore, as long as the current trip threshold for each port is set below the 5.0 Amp UL limit no other threshold needs to be maintained. When ganging the ports this particular point becomes a bit more important.

Ganged vs. Non-Ganged Ports

In the case of the high-powered port output, up to 500 mA of continuous current can be drawn from the port. Typical systems may have up to four ports and even as many as seven. With seven ports ganged through a single current limiting device, you could draw a total of 3.5 amps through it.

There are several drawbacks to a ganged configuration. The first is that the design limits are very close to the UL current limit. This can make it difficult for the overcurrent protection device to distinguish between a fault and normal operation in some situations. Another limitation is that a single fault can shut down all of the ports ganged through the same device. When implementing any type of ganged configuration, one should consider limiting it to three or four ports per device, for operational purposes.

A ganged output provides the most economical solution to the voltage distribution requirements of USB, but this also requires the use of higher current devices. Individual or non-ganged solutions can provide greater system reliability and versatility. The overcurrent protection device for an individual port must only meet the need of that particular port, which is less than 500 mA. A ganged output will always, at minimum, require a current limit, which is a multiple of the 500 mA plus some tolerance, with a 5 Amp maximum.

Another consideration in a ganged approach is the effect on the shared voltage bus during hot plug/unplug of all of the peripherals. The voltage drop and droop regulation requirements must be maintained regardless of the system configuration. Depending on the ganging configuration, an action on any single port may dramatically affect the remaining ports. Hot plugging generates a voltage droop, which can lower the bus voltage to below the voltage required for devices to operate properly. Ganging ports alters the voltage droop of the system and requires lower resistance OCP devices.

USB Power Application Notes

The basic overcurrent requirement of the USB specification calls for the recognition of catastrophic device failures or shorts. The specification does not require the host to recognize illegal topologies or excessive loading. It is quite possible for the system to be configured in such a manner that the current drawn through a port is well in excess of the 100 mA or 500 mA allotted, but still beneath the overcurrent trip threshold. With the implementation of some type of thermal limit, the device may be able to recognize excessive current draw. The device could then limit port current to some nominal current level or shut down the port, without reaching the current trip point.

The USB specification does not require overcurrent detection on low-power ports. USB only requires a switch on these voltage outputs. The overcurrent condition monitoring takes place in the host device in the USB system. The switches allow the bus-powered hub to shut down its output ports and power up with a maximum current draw of 100 mA. The downstream ports will receive power after the hub has been recognized and configured by the system. Switches with controlled turn-on times effectively limit the inrush current as the hub and each subsequent port is powered-up. Controlling inrush current (soft start) helps limit damage to system devices and is required by the specification.

Although not required, some type of current limit may be advisable in the bus-powered hubs. The hub controller may manage this power interface and can communicate a condition back the host, in some cases. Even without the reporting aspect, allowing these ports to be current limited can add to the integrity of the system.

8.1.5 Connectivity Limitations

Due to the voltage drop in connectors, contacts, p.c. board traces, cables, and circuit devices the voltage budget limits the bus to two tiers. The two tiers are: the Host/SP-Hub output to a BP hub, and BP-Hub output to a function. Section 7 of the USB specification shows this voltage limitation, also shown on this page. Working self-powered hubs into the system properly allows this string of

Chapter 8: Appendix

peripherals to contain up to 127 devices (not more than seven rungs deep, due to signal delay considerations).

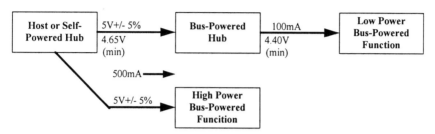

The figure shows the voltage limitations of a typical system configuration. The 4.40 V out of the bus-powered hub drops 250 mV in the cable before reaching the low-powered function. Most USB function designs require a 3.3 V operating voltage, and with standard low dropout regulators this would only leave 85 mV dropout voltage for the device. The voltage out of the last device would not be large enough to allow any connections to additional devices at current operating voltages.

8.1.6 Power Supply Voltages

USB is a system that distributes 5 V power to run the USB Hubs and functions in the system. The USB tolerances allow the voltage bus to drop to a minimum of 4.65 V. This makes it quite clear that the USB controllers, hubs, and functions powered by the bus must be designed to run on a voltage significantly lower than 5 V.

Low drop-out voltage (LDO) regulators implemented in the controller power supply sections of USB devices will allow these peripherals to function at lower bus voltages. Typically a Hub controller design implements a 3.3 V supply, because data travels on the USB bus at 3.3 V logic levels. The minimum operating voltage of the hub controllers and functions determines how low the drop-out voltage on the LDO has to be.

Given a 4.65 V minimum out of the host or bus-powered hub, and a possible 350 mV voltage drop before reaching the regulator, the bus-powered hub would require an LDO with a drop-out voltage of less than 1 V. This same situation exists in the functions connected to the hub, but the drop-out voltage required is much lower. The end-of-line (EOL) function could see input voltages as low as 4.1 V given a 4.40 V minimum output voltage from the Hub. This would require the LDO drop-out voltage to be in the neighborhood of 800 mV, if the function operates at 3.3 V.

These voltage limits do not take into consideration the voltage droop seen when hot-plugging other devices on the system. The USB specification allows a 330 mV voltage droop when hot-plugging occurs. Due to bulk capacitance in the system, the peripheral device should only see a fraction of this voltage droop, but even that could be enough to fall below the operating voltage limit of the LDO.

Designs should consider all of the system loading possibilities, and each device should be designed to keep system loading during hot-plugging to a minimum. A typical hub can draw as much as 100 mA for the controller, and may draw as much as 500 mA to power the controller and function combined. The hub can distribute any portion of the 500 mA bus current it does not use, in 100 mA increments. The distributed current not consumed by the hub does not pass through the LDO.

At minimum, a USB device requires a 100 mA LDO with a drop-out voltage less than 800 mV. In a 500 mA application, the allowable drop-out voltage only moves up to approximately 1 V. Typical designs use devices that meet requirements at least twice that which is required. With this in mind, the 100 mA LDO with the 800 mA drop-out may be specified at 200 mA with a 650 mA drop-out voltage, and the 500 mA device may move to 1 A with a 750 mA drop-out voltage. In these situations it becomes the peripheral and hub designer's job to work around the worst case scenarios.

8.1.7 Voltage Requirements

The voltage requirements of the USB interface are quite specific and directly relate to the voltage drop that the current limit device may add to the system. The voltage drop requirements define specific on-resistance limits for switching devices.

The 5% tolerance of the 5 V power supply allows the voltage out of the host or self powered hub to float between 5.25 V and 4.75 V, with a minimum host output set at 4.65 V. There is a minimum of 4.40 V output voltage for the downstream ports from a bus-powered hub. Due to device limitations, the low power functions need to be fully operational at this voltage level minus the voltage lost in the system connections.

According to Section 6.4 of the USB specification, the voltage drops at 500 mA across the various loads on a typical bus are:

Board resistance and switch = 100 mV

Connector contacts (each) = 15 mV

Cable conductors (each) = 95 mV

The system power requirements are:

Output Voltage Regulation = ± 5%

Power Supply Reg. (typ) = ± 3%

(4.65 V output allows for 5%)

Evaluating all of these parameters allow calculation of a specific on-resistance, for the applicable current limit device. Examining the IR drops on a systems level, based on the hub interconnect constraints, yields the following formula:

$$V_{HUB} \geq V_{SWITCH} + 4*V_{CONNECTOR} + 2*V_{CABLE}$$

The voltage drop across the switch and p.c. board traces is defined as 100 mV, and the resistance calculation reduces to:

$$R_{ONS} \leq V_{SWITCH}/I_{MAX} * N$$

Where N = # of ports ganged through a single OCP device.

$$R_{ONS} \leq 100mV/500mA$$

For a value of N=1 (single non-ganged port).

Therefore:
$$R_{ONS} < 200 \text{ m}\Omega$$

This is a system on-resistance (R_{ONS}) expectation for one port and includes the resistance that the p.c. board traces add to the system. In order to remove the p.c. board resistance from this on-resistance, the voltage tolerance on the self-powered hub must be evaluated. Based on typical system requirements, the following formula is generated for the value of the regulated output voltage (V_{REG}):

$$V_{REG} > V_{PSREG} + V_{PCB} + (I_{Omax} * R_{ON})$$

The value $I_o * R_{ON}$ is equivalent to V_{DEVICE}, voltage drop across the device, for certain types of components. Solving this equation for the switch on resistance:

$$R_{ON} \leq \frac{V_{REG} - V_{PSREG} - V_{PCB}}{I_{Omax}}$$

When the voltage values for the output regulation, the regulated supply, the p.c. board voltage drop, and the 500 mA hub current (a single non-ganged port) are put in, the calculation yields:

$$R_{ON} \leq 150 \text{m}\Omega$$

Since this (R_{ON}) is the lower of the two resistance values calculated and it does not include the resistance of the p.c. board, this value is the maximum allowable on- resistance for the current limit device for a high powered port. This value would be lower by a factor of N for N ganged ports. A low-power load is only 100 mA, and the allowable resistance is five times that allowed for the high-power load at 500 mA.

The current limit device in the host and self-powered hub must meet the same rigorous resistance requirements as the switching device used on the voltage bus in the bus-powered hubs. The additional 100 mV drop at the 4.65 V output allows for power supply regulation to move from 3% to ±5%. The additional drop allowed in the power supply does not necessarily add to the voltage drop allowed in the current limit device.

8.1.8 Current Limiting Devices and Power Switches

As stated in the USB specification, the current limit device for the self-powered hub or host can include but is not limited to Polyfusts, fuses, and solid state switches. Consider the following USB requirements, as well as typical system requirements, for the USB voltage bus prior to selecting a current limit device:

- Power to downstream ports must be current limited to less than 5 A
- Power to downstream ports must be switched in bus-powered hubs
- An overcurrent response must be reported to the host USB controller
- Incorporating thermal protection into the overcurrent protection is recommended
- The system voltage budget from upstream cable connector to downstream hub connector is 250 mV
- Maximum allowable switch/device resistance due to voltage tolerances is 150mΩ for a 500 mA port

Given these requirements, a review of the possible overcurrent protection devices shows the following plausible solutions:

1. Positive Temperature Coefficient (PTC) Resistors (i.e., Polyfuse, PolySwitch):

- Cannot be controlled (switched) by the host or USB controller
- Cannot report an overcurrent condition to the controller
- Cannot maintain an acceptably low on-resistance.

USB Power Application Notes

Examples: (Raychem PolySwitches)

RXE110　　　Hold Current = 1.10 A

　　　　　　Trip Current = 2.20 A

　　　　　　Resistance, Initial = 0.2 Ω

　　　　　　Post-trip Res. = .38 Ω

SMD050　　　Hold Current = 0.50 A

　　　　　　Trip Current = 1.0 A

　　　　　　Initial Resistance = .35 Ω

　　　　　　Post Reflow Res. = 1.4 Ω

Considering the specifications of these devices, and the requirements of USB, polyfuses are not acceptable current limit devices for the USB power interface.

The resistance of a polyfuse is too high at the expected current levels to maintain the proposed voltage budget. Polyfuses will not maintain the required regulation on the voltage bus unless a device with constant current value in excess of two amps is used; the device's inherent resistance is too large below this point. Each time a polyfuse trips the resistance value of the fuse increases. The higher valued polyfuses have longer trip times and may cause wide voltage swings on the voltage bus during a fault before reacting.

2. Fuses:

　　Cannot be controlled (switched) by the host or USB controller

　　Cannot report an overcurrent condition to the controller

　　Require additional maintenance from the end user or service center

Although fuses are an excellent OCP, they are not a practical interface due to the maintenance required. The resistance of the fuses will normally be acceptable for the application, but this solution may require continual maintenance. Like polyfuses, they do not provide feedback, and they are not a controlled switch.

3. Solid State Switches: (Digital Relays)

    Cannot report an overcurrent condition to the controller

    Generally do not have integrated overcurrent protection

    Do not have controlled on-time for inrush current limiting

Solid state switches, like those from CotoWabash or Teledyne, have a digital control interface but do not monitor current. They can be used in conjunction with a current limiting device to provide a controlled (switchable) power interface, but will still not provide the required overcurrent feedback to the host controller.

4. Power Distribution - MOSFET switches:

    Provide integrated current limit

    Provide interface for host controller

    Provide integrated thermal sensing

    Provide overcurrent response

    Provide controlled switching times

    Provide on-resistances well below the 150 mΩ max.

By far, an intelligent MOSFET switch provides the best USB power interface on the market today. Typical MOSFET switches have on resistances ($r_{DS(on)}$) ranging from 50 to 150 mΩ. A 50 mΩ switch would allow ganging of two or three 500 mA ports and still maintain the required system voltage regulation.

Intelligent MOSFET switches similar to the Texas Instruments TPS20XX, Micrel MIC250X, Maxim MAX890, and Linear Tech LTC1477 devices provide feature sets applicable to the USB power interface device. Each family of devices provides low on-resistance and current limiting. Some provide overcurrent response capabilities, thermal limiting, and a digital interface for switching control. Not only is a MOSFET switch a perfect implementation for the host- and self-powered hubs, the MOSFET switch is also ideal for the bus-powered hubs where only a switch and inrush current limiting is required.

In a bus-powered hub where additional functionality is not required, these MOSFETs still function as on/off switches, with a very low on-resistance. This allows for the same device to be used throughout the system as a USB power interface.

8.1.9 Conclusion

The USB specification sets up the requirements for the design of the power-distribution interface for the voltage bus. The specification even mentions that fuses are a functional device for this interface. However, simple implementation, reset without maintenance, high reliability, long life, thermal limited, integrated switch, and low cost should be the general design guidelines for selecting a USB power distribution current limit device. Fuses do not have all of these characteristics.

The current protection device's characteristics should include on resistance less than 150 mΩ, overcurrent trip below 5 A, current limit below 2.5 A, digital interface (controllable), and an integrated overcurrent response. The only devices that typically meet all of these requirements are MOSFET switches. A MOSFET switch with current limiting, fault indication, and thermal protection is the only integrated solution currently available that meets all of the requirements of the USB specification.

8.1.10 References

USB Specification: Version 1.0

Review Request 94: Intel/USB forum

8.2 USB Power Management

This section is from a technical paper by Kevin Lynn PE, USB Applications Engineer at Micrel Semiconductor. Permission to reprint this material is gratefully acknowledged. Questions to Mr. Lynn may be directed to him through klynn@wco.com. Additional product information is available at http://www.micrel.com.

8.2.1 Abstract

Universal Serial Bus (USB) is a new peripheral standard that will bring Plug and Play capability to peripherals outside the computer cabinet, removing the user's need to open the case to install cards into physical slots, or to understand and set interrupts or device addresses, or to troubleshoot resource conflicts.

Power management and distribution is a major factor in correctly designing USB peripherals. Proper methods of designing USB peripheral power distribution are crucial to ensure full compliance with the USB specification, including satisfaction of EMI and voltage droop requirements.

Criteria for circuit board layout, component selection, steady state and dynamic voltage levels, and EMI control are described. Power control elements include linear voltage regulators, resettable polyfuses and integrated high-side power switches. Footnotes show paragraph references to USB Specification, version 1and recent revisions.

Calculations are shown for determining worst-case voltage drop and transient droop during hot-connect. Circuit examples are given for simple and complex Self-Powered and Bus-Powered USB hubs.

8.2.2 Introduction

Universal Serial Bus (USB) is a new peripheral standard that will bring Plug and Play capability to peripherals outside the computer cabinet, removing the user's need to open the case to install cards into physical slots, or to understand and set interrupts or device addresses, or to troubleshoot resource conflicts.

USB experienced phenomenal growth in 1997, and is expected to become pervasive in PCs and peripherals in 1998. The bus provides 1.5 Mbps and 12 Mbps two-way serial data plus 5V operating power over pluggable four-wire cables to peripherals such as keyboard, modem and mouse. The USB Implementers Forum (USB-IF) controls a specification detailing mechanical, electrical, software and signal requirements. The USB Device Working Group (DWG) is made up of manufacturers, recommends changes to the specification. The USB specification will be updated to revision 1.1 during 1998, incorporating most of the DWG agreed changes, where devices that comply with revision 1.0 are not disabled.

8.2.3 Power management

Power management and distribution is a major factor in correctly designing USB peripherals. Proper methods of designing USB peripheral power distribution are crucial to ensure full compliance with the USB specification, including satisfaction of EMI and voltage regulation requirements.

Unit Load

The power sourcing and sinking requirements of different USB device classes may be simplified with the concept of a unit load. A unit load is defined to be 100mA.[1] "The unit load a device can draw is an absolute maximum, not an average over time. The load is measured at the upstream (toward the host) end of attached cables or at the Series B connector pins of a device that allows a standard USB cable".[2] The unit load is equivalent to 44 Ω in parallel with 10 µF. At the downstream (away from the host) connector of a Bus-Powered hub, a unit load will draw 100mA at 4.40V minimum

---

[1] USB Specification, Rev. 1, ¶ 7.2.1
[2] USB Specification, Rev. 1.1, ¶ 7.2.1 , DWG Review Request 97, agreed 10/10/96

voltage. In this paper, "100mA" means one unit load, and "500mA" means five unit loads, even though the actual current may be higher due to applied voltage.

8.2.4 Device Classes

USB defines several classes of devices:

Hubs

Hubs have a USB controller, distribute data and power downstream and communicate with an upstream host. Hubs may be self-powered or Bus-Powered, or a combination of the two.[3] Bus-Powered hubs must provide port power switching; Self-Powered hubs must provide overcurrent protection at the ports.[4]

Host

A USB system may only have one host, the highest device upstream[5]. A host may not draw power from downstream, so must be Self-Powered, but is not required to provide more than 100mA downstream power.

Self-Powered Devices

Self-powered devices include hubs and functions. Internal operating power is obtained from the upstream source over the bus cable, limited to 100mA, or from a local supply.

---

[3] USB Specification, Rev. 1, ¶ 11.7
[4] USB Specification, Rev. 1.1, ¶ 11.7 , DWG Review Request 104, agreed 10/10/96
[5] USB Specification, Rev. 1, ¶ 10.1

USB Power Management

Self-Powered Hub

A self-powered hub repeats data, may supply up to 500mA to each of its downstream ports[6], with an overcurrent limit of 25VA per port.

Self-Powered Function

The self-powered function's controller may be powered from the upstream cable or from a local power supply. When externally powered, the maximum power drain from upstream is 100mA, with inrush current limiting. The amount of total power drawn is limited only by the local power supply. Since a Self-Powered function has no downstream ports, USB does not require it to have current limiting, soft start or power switching.[7]

Bus-Powered Devices

Bus-powered devices include hubs, high power functions, and low power functions. Bus-Powered devices draw all of their power from the upstream cable. They may draw 100mA on power-up, and up to 500mA, split between any embedded functions and external ports. High power functions, such as a modem, may draw up to 500mA from an upstream source, over the bus cable. If more power is required, then the device must be self-powered.[8] Low power functions, such as a mouse, draw up to 100mA on attached upstream cables.

High-power Function

High-power functions, such as a modem, may draw from 100mA to 500mA from an upstream source, over the bus cable. A high power function requires staged switching of power, initially drawing

---

[6] USB Specification, Rev. 1, ¶ 7.2.1.2
[7] USB Specification, Rev. 1, ¶ 7.2.1.5
[8] USB Specification, Rev. 1, ¶ 7.2.1.1

100mA. They must initially operate on a minimum of 4.40V, measured at the upstream connector.

If sufficient power is available from the upstream hub, additional power may be drawn, up to 500mA. If additional power is needed, a local power supply is required. They must be able to operate at full power (up to 500mA) at a minimum of 4.75V measured at the upstream end of the cable. If there is insufficient power, the remainder of the function is not powered, and a power limit warning message is sent.[9]

Low-power Function

Low-power functions, such as a mouse, may draw up to 100mA on attached upstream cables. They must operate on a minimum of 4.40V, measured at the upstream connector.[10]

8.2.5 USB Power Distribution

Physical connection between devices is made with standard length detachable or various length attached shielded high-speed or unshielded low-speed four-wire cables, providing 5V power and differential isochronous data to downstream hubs and peripherals.

There are two major types of interconnect hubs, which supply power downstream and two-way communications to the upstream host. Hubs may be either locally powered (Self-Powered) or powered from the upstream cable (Bus-Powered), or a combination of the two. Downstream port power switching may be on a per port basis or have a single switch for all of the ports (gang mode power control).[11]

---

[9] USB Specification, Rev. 1, ¶ 7.2.1.4
[10] USB Specification, Rev. 1, ¶ 7.2.1.3
[11] USB Specification, Rev. 1.1, ¶ 11.7 DWG Review Request 104, agreed 10/10/96

Self-Powered Hubs

Self-powered hubs (SPH) have a local power supply, such as PC's with USB ports, stand-alone hubs, monitors, printers, scanners, and docking stations, drawing up to 100mA (a unit load) from an upstream port. SPH may supply up to 500mA to each of up to seven downstream ports.

Power Supply Ground Isolation

The chassis ground (if one exists) of a Self-Powered Device or Function should be DC isolated from the USB signal ground, to prevent ground loop current on the cables.[12] "The complete bus should have only one DC ground point at the host end."[13] Local power supplies may provide ground isolation with an isolated transformer winding for bus power, connecting the negative power lead to the ground plane of the hub, with no "green wire" metallic connection to earth ground, except at the host. (This requirement may be relaxed to "recommended" in Rev. 1.1).

Monitor

The video display (monitor) is a good candidate for installation of an SPH. Monitors are normally located in a central location, near the user, keyboard, and mouse. Monitors have an internal power supply which can be adapted to provide the under 25VA maximum power drain of an SPH. A monitor's video cable normally provides a chassis ground connection upstream, so the chassis ground of a hub supply in a monitor may also be grounded, if both power cords use the same outlet.

---

[12] USB Specification, Rev. 1, ¶ 7.2.1.2.2
[13] USB Specification, Rev. 1, ¶ 6.6

Printer

A printer is also a good candidate for an SPH, providing communications and power to peripherals such as a scanner, modem or other peripheral not required to be located near the user.

Docking Station (laptop)

Laptop computers have limited power when operating from their internal battery. When attached to a docking station, the local power supply may also provide power for an SPH.

Stand-Alone Self-Powered Hub

Many users will want to add USB peripherals to a system that has a non-USB monitor or printer. A stand-alone SPH can provide regulated power and connectivity upstream to the host.

8.2.6 Self-Powered Hub Requirements

A local power supply provides 4.85V to 5.25V maximum at loads up to 500mA per port. USB requires a minimum of 4.65V at the downstream output connector of a host or self-powered hub (SPH). SPH may draw up to 100mA operating power from an upstream connector, with an initial plug-in equivalent load of 44Ω in parallel with 10μF. Each downstream port is limited by the specification to a maximum of 500mA, with a short-circuit maximum of 25VA from any port. A practical limit is seven downstream ports (3.5A) per protection device.

Hub Power Supply Separation

The USB power wire (Vbus) from an upstream port in a self-powered hub must be separated from the downstream wire (Vbus), so that current cannot flow upstream.[14] Power for the hub controller

---

[14] USB Specification, Rev. 1, ¶ 11.7

may be supplied from either the upstream port (a "hybrid" powered hub) or the local power supply. If the hub is powered from the upstream cable, communication with the host is possible even if local power is disconnected. "If power is supplied from the bus, it can draw no more than one unit of current to allow it to be attached to and operated from any port."[15]

Overcurrent Protection

"For reasons of safety, all locally Powered hubs must implement current limiting on their downstream ports. Under no conditions may more than 25 VA be drawn from any USB hub port. (The actual overcurrent trip points may be lower than this value. If an overcurrent condition occurs, even if it is only momentary, it must be reported to the hub controller. ... Detection of overcurrent must disable all affected ports. If the overcurrent condition has caused a permanent disconnect of power (such as a blown fuse), the hub must report it upon coming out of reset or power-up. Overcurrent protection may be implemented over all downstream ports in aggregate, or on a per port basis. The ports may optionally be split into two or more subgroups, each with its own overcurrent protection circuit."[16]

Overcurrent Limiting Devices

Overcurrent limiting devices may include polyfuses, standard fuses or a solid state switch. Current limiting should not occur even if illegal topologies are configured, since high-power functions and BPH have power switches that will disable illegal loads.[17] Overcurrent circuits are intended to prevent catastrophic device failures, such as short-circuited connectors or software power-control errors.

---

[15] USB Specification, Rev. 1.1, ¶7.2.1.2 revised by Review Request 93, approved 12/3/96
[16] USB Specification, Rev. 1, ¶ 11.7.1
[17] USB Specification, Rev. 1, ¶ 7.2.1.2.1

Chapter 8: Appendix

Resettable Polymer Positive Temperature Coefficient Fuses

A self-resetting 2.5A PPTC resettable fuse (polyfuse) may take up to 100 seconds to trip at 5A.

Polyfuses interrupt current by undergoing a resistance increase when heated above a threshold, limiting the output current. This resistance increase initiates thermal runaway, rapidly heating the device and increasing its resistance. After an overload is removed, normal current flow resumes through increased resistance continues the self-heating, and the on-resistance of a device is up to four times its cold resistance an hour after a trip event. Polyfuses permanently increase their resistance after a trip event, or exposure to high temperature, such as wave soldering.

PPTC Continuous Current Rating

A typical 2.5A polyfuse – specified to trip at 5A - will pass a continuous 2.5A, with a never tripped on-resistance ranging from 30mΩ to 50mΩ at 25°C ambient, rising to approximately 70mΩ after its initial trip.

Ganged Polyfuse Trip Time

In order to supply 500mA to each of the seven ports of an SPH, two 2.5A polyfuses are needed, feeding up to four ports each (ganged protection).[18] At 40°C ambient, the typical 2.5A polyfuse continuous current rating drops to 2.1A, while its resistance rises to 130mΩ, taking up to 100 seconds to initially trip at 5A. When tripped the resistance becomes very high.

Individual Polyfuse Trip Time

In order to reduce trip time, a separate 0.5A polyfuse could be placed on each output line. A typical 0.5 A polyfuse has 1.2Ω post-trip on-resistance, providing 5A trip times of 0.2s, producing a

---

[18] USB Specification, Rev. 1, ¶ 11.7.1

voltage drop of 600mV at normal 500mA hub load, requiring an input supply voltage above 5.25V, violating the USB specification for 100mV maximum drop.

Power Regulators

Linear Regulators provide a cost-effective method of supplying closely regulated standard voltages from unregulated power supplies. Low DropOut (LDO) regulators, such as Micrel's 3 Amp MIC29301-5.0 provide current limited +/-2% regulated output over a wide range of input voltage and output loads. This LDO has a logic enable pin and an error flag output that indicates current limiting or over-temperature. 5V LDO regulators can provide regulated voltage output with as little as 5.35V input under full load.

Current Limit

The MIC2930x LDO is rated for continuous 3A output, with a typical 3.8A current limit, short-circuit limit less than 5A.

Output Enable Pin

The MIC29301 and MIC29302 have a logic-level non-inverting output enable pin, allowing the output to be turned on or off.

Error Flag

The MIC29301 and MIC29303 LDOs have an open-drain error flag that conducts to ground if the device is in current limit, over-temperature shutdown, or low input voltage (excessive dropout).

Discrete Switches

Discrete MOSFET switches require additional components to be controlled by a logic signal. N-channel FETs need a high-side driver to boost the gate to source voltage above the power supply by at least 5V. P-channel FETs require a pull-up resistor from gate to drain, and level-shifting logic to pull the gate down by at least 5V to turn the switch on. Discrete switch devices do not have inherent current limiting, soft start or over-temperature shutdown.

8.2.7 Integrated High-Side Power Switches

Integrated high-side power switches, such as Micrel's MIC2525 single and MIC2526 dual switches react quickly to overloads, limiting current to <1.5A, providing better safety margin and overall power consumption control, and have an error flag output that signals overcurrent, over-temperature, or undervoltage.

Power Switch Selection

Table 1 shows characteristics of Micrel High-side Power Switch products.

Table 1. Micrel Power Switches

Part	Switches	I_{out}	R_{on}
MIC2505	single	2.0A	35mΩ
MIC2506	dual	1.0A	70mΩ
MIC2525	single	0.5A	140mΩ
MIC2526	dual	0.5A	140mΩ
MIC2527	Quad	0.5A	140mΩ

Switch Enable (V_{on}/V_{off}) is available as inverting or non-inverting logic, allowing flexibility in choice of hub controllers. All integrated switches are guaranteed to survive an output short-circuit with a 5.25V input.

Voltage Drop

A 100mV drop is typical across the host or self-powered hub circuit board to each downstream connector, including the input connector from the power supply, printed circuit board resistance and the on-resistance of the overcurrent protection device. The DC output voltage, measured at the board side of the SPH downstream connector, must remain above 4.65V under all legal continuous load conditions.[19] Voltage drop is measured beginning at the local power supply connection to the board, 5V, +5%, -3%. Figure 1 shows the drops along one 0.5A trace in a typical SPH. Functions drawing more than 100mA must operate with a minimum of 4.75V at their upstream connector.

8.2.8 Transient Droop

Ferrite beads reduce electromagnetic interference (EMI), and also limit the inrush current during hot-attach. Ferrite beads act to attenuate high frequency signals while allowing DC power to pass freely. The simplest ferrite beads consist of a small ferrite tube on a solid tinned copper wire, or surface-mount equivalent. The resistance of the ferrite bead wire should be as low as possible, with a large solder pad to minimize connection resistance. Beads placed on the 5V power line should be tested for conduction to ground, or have insulation, to prevent leakage current flow through the ferrite. USB requires that any momentary droop never cause the 4.75V at a BPH's upstream cable connector to droop more than 330mV to below 4.42V. Likewise, the voltage transient at a 4.40V peripheral's upstream cable shall never go below 4.07V.[20] The worst case transient droop will occur with connection of a device with a very short cable and 10µF, such as an IrDA "dongle". Good layout, with wide, short traces and a minimum of 33µF Tantalum or 100µF Electrolytic capacitance per port (120µF per hub) will prevent droop from exceeding 330mV.

8.2.9 Layout

The printed circuit board power and ground traces, solder connections and resistance of the ferrite beads on both power and ground output lines may total 60 mΩ, dropping a total of 30mV at 0.5A. The remaining voltage left from the 0.1V overall drop budget

---

[19] USB Specification, Rev. 1.1, ¶ 7.2.2, revised by Review Request 94, approved 8/28/96
[20] USB Specification, Rev. 1, ¶ 7.2.4.1

is 70mV, which sets the maximum resistance allowed for an overload protection device at 0.14Ω. Additional traces, ferrite beads and protection devices are needed for each 500mA output port.

Thermal Management

Power conduction causes heat in circuit resistance, which must be dissipated to prevent thermal problems. Surface-mount packages have limited heat dissipation, due to their small size. The traces connected to packages, and surface area of nearby copper, provide a major heat sink for power dissipated inside the package.

Power Dissipation

Heat is produced by current flow (I) through resistance (R). Power dissipation (Pd) in integrated circuits causes a temperature rise above the ambient temperature (Ta). An SO-8 surface-mount IC package (BM) has a thermal resistance (θja) of 160-180°C/W:

$$Pd = Ta + I^2R*\theta ja$$

Current flow produces a temperature rise above ambient at the junction of the die and the package. A minimum of one square inch of copper near or under each package is recommended to remove the added heat. Use of a thermally conductive substance, such as epoxy paste, between the package and the circuit board, aids heat transfer.

Table 2. Die Temperature vs. Current
θja = 160 deg°C/W

MIC2526BM	0.1	0.25	0.5	1	Amps	
Ambient °C	0.001	0.01	0.03	0.10	W	Ron
25	25.2	26.0	29.0	41.0	deg°C	0.100
50	50.2	51.1	54.4	67.6	deg°C	0.110
75	75.2	76.2	79.8	94.2	deg°C	0.120
100	100.2	101.3	105.2	120.8	deg°C	0.130
125	125.2	126.4	130.6	O.T.	deg°C	0.140

USB Power Management

Table 2 shows the junction temperature of an MIC2526BM SO-8 package at various ambient temperatures for different currents in one of the dual power switches.

An 8-lead plastic dual-inline package (8 PDIP) has a thermal resistance (θja) of 100-110°C/W, with approximately five times the mass of an SO-8, and four times the radiating area.

Thermal Overload

Thermal overload protection is actuated when an MIC2526 junction rises above 135°C, turning off the output of both switches (shown in table 2 as O.T.), until the die cools down to 125°C. A slow (1-5Hz) on-off oscillation - turning the output off when the junction reaches 135°C, and turning it back on when the die cools to 125°C - will occur as long as the output load is heavy enough to force the device into thermal shutdown. By sensing the voltage fall to zero when over-temperature limiting begins, the USB hub controller may turn off the switch Enable pin, preventing an on-off oscillation. A timer function may be implemented in the controller to periodically re-enable the switch to determine if the overload has been removed.

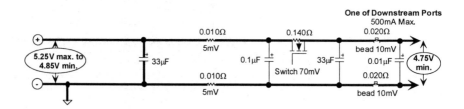

Figure 1. Self-Powered Hub Voltage Drops

Self-Powered Hub Voltage drops

Figure 1 shows the various voltage drops caused by printed circuit trace resistance, solder joints, power switches, and ferrite beads on power and ground leads for an SPH.

Voltage Measurements

All voltages are measured from connector power pin to connector ground (return) pin. Traces which carry the combined current from the input should be made heavier to minimize voltage drops, or separate wide traces should be laid out directly from the input filter capacitor to each switch input pin.

Voltage Drop Budget

Adding the drops from the minimum output voltage of 4.75V shows that a minimum input of 4.85V is required to ensure adequate output. The overall voltage drop to each port, caused by the printed circuit board and overcurrent protection device at full seven-port 3.5A load current, is typically less than 100mV. Ground traces are as important as positive traces, as all voltage drops are in series.

Voltage Drop Calculations

Voltage drop is developed by current flow through resistance: V=IR. The voltage drop across an SPH board has three components: board (Vpc), protection (Vp), and output filter (Vfb).

Board Voltage Drop

Each output port of a self-powered hub is required to supply up to 500mA, so the voltage drop in the PC board (Vpc) is the trace resistance of the power and ground paths, approximately 15mΩ each, times 0.5A;

$$Vpc = 2 * 0.015\Omega * 0.5A = 15mV.$$

Output Filter Voltage Drop

Each output port of an SPH is required to supply up to 500mA, so the voltage drop in the output filter (Vfb) is the resistance of two ferrite beads and their solder joints, approximately 15mΩ each, times 0.5A;

$$Vfb = 2 * 0.015Ω * 0.5A = 15mV.$$

Protection Device Voltage Drop

Each output port of an SPH is required to supply up to 500mA, so the voltage drop in the protection device (Vp) is the resistance of the device and its solder joints. The sum of the voltage drops from the other components is:

$$Vpc + Vfb = 15mV + 15mV = 30mV.$$

The recommended voltage drop allowed across an SPH is 100mV, so the loss in the protection device (Vp) cannot exceed 70mV:

$$Vp = 100mV - 30mV = 70mV.$$

At an output port current of 500mA, the protection device's resistance may be up to 140mΩ:

$$Vp = 0.14Ω * 0.5A = 70mV.$$

Total Self-Powered Hub Voltage Drop

The total voltage drop across the SPH board, protection device and output filter is:

$$Vpc + Vp + Vfb = 15mV + 15mV + 70mv = 100mV.$$

Subtracting the total drop from the minimum input voltage of 4.85V leaves 4.75V between the power and ground pins of each downstream SPH port.

Chapter 8: Appendix

Unregulated input Self-Powered Hub Example

Figure 2 shows a minimal self-powered hub, using Micrel's MIC29301-5.0BT Low Drop-Out (LDO) 3A linear regulator to provide 5.0V +/- 2% (4.9 to 5.1V) from an unregulated >5.35V, 3.5A transformer-rectifier-capacitor input.

Component Selection

The MIC29301-5.0BT has an enable pin to control ganged power up to 3.5A, to seven downstream ports, each drawing up to 500mA. A four-port SPH may use a 1.5A MIC29151-5.0BT LDO in place of the higher current MIC29301.

Protection

MIC29301 limits current to <3.8A, with <5A short-circuit current. An open-drain error flag goes low if the device is in current limit, thermal shutdown or has an excessive output voltage drop. If heavy loads cause the over-temperature shutdown to activate, a slow on-off oscillation may occur as the regulator heats up under load and cools down when shut-off. The hub controller may be programmed to shut down the LDO after the error flag has been low for a preset time, periodically re-energizing the LDO enable to determine if the excessive load has been removed.

USB Power Management

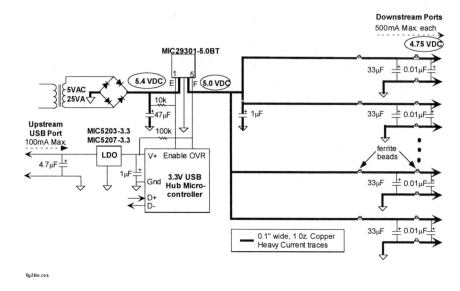

Figure 2. Stand-Alone or Unregulated Input Self-Powered Hub

Output Filter

Output ports must have bulk 120µF, 16V tantalum capacitors[21] across the power and ground lines, near each downstream connector, to reduce EMI and decouple voltage droop caused by hot-plug of downstream cables. Ferrite beads[22] in series with the V_{BUS} and Ground lines, and 0.1µF bypass capacitors[23] at the power connector pins, are recommended for EMI and ESD reduction.[24] A four-port hub with a 33µF, 16V tantalum or 100µF, 10V Electrolytic capacitor per port should meet the 330mV maximum voltage droop requirement.

Regulated Input Self-Powered Hub Example

With local 5V +/-5% power supply – e.g., a standard PC "silver box" - providing operating voltage from 4.85 to 5.25V, protection

---

[21] USB Specification, Rev. 1, ¶ 7.2.4.1
[22] USB hub checklist
[23] USB Specification, Rev. 1, ¶ 7.2.4.2
[24] USB Specification, Rev. 1, ¶ 7.2.1.2.2

Chapter 8: Appendix

devices with less voltage drop than an LDO regulator are needed. Micrel's Integrated High-Side Power switches have low on-resistance, with built-in current limiting and logic-level enable. The worst-case voltage drop recommended across the SPH board, MIC2526 and ferrite beads is 100mV, providing 4.75V to an output port.

Component Selection

Figure 3 illustrates an SPH with up to seven downstream 500mA switched ports, using multiple MIC2526 dual 0.5A power switches.

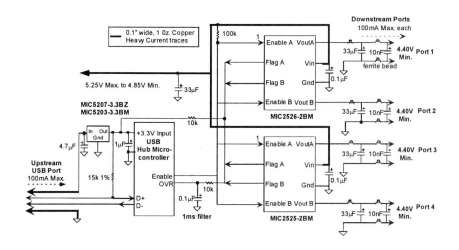

Figure 3. Self-Powered Hub with Individual Power Switches

Protection

Micrel's MIC2525 single and MIC2526 dual 0.5A switches have soft start, ramping up turn-on current over 1-2ms, short-circuit protection, with current limit, under-voltage and over-temperature flags. The hub controller monitors the downstream status. Enable pins on each switch section allow the hub controller to shed loads if needed to reduce current drain.

USB Power Management

Voltage Drop

The switches have maximum switch on-resistance of 140mΩ with a 500mA load, producing a maximum voltage drop of 70mV, leaving a balance of 30mV to circuit board trace resistance to meet the SPH 100mV recommended overall drop. Soft start provides inrush current limiting, minimizing voltage droop to other ports when any switch is enabled.

Other Features

Traces carrying power current are made wide to reduce resistance, with each switch having a separate Vin trace from a 33µF bulk capacitor.

Output Filter

120µF tantalum capacitors across each output, with ferrite beads in series with downstream Vbus and ground lines, prevent voltage and current surges caused by hot-attach of downstream devices from disrupting power to other attached devices.

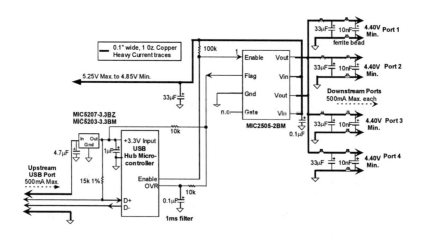

Figure 4. Ganged Switch Four Port Self-Powered Hub

Chapter 8: Appendix

Regulated Input Ganged Switch Self-Powered Hub

Figure 4 shows a four-port SPH, using a single MIC2505 2A power switch to provide ganged power from a 5V +5% -3% regulated local supply, with three ferrite beads and a 33µF, 16V tantalum capacitor on each port.

EMI control and Transient Voltage

Whenever current is switched rapidly, electromagnetic interference (EMI) may be produced, disrupting communications, video, radio, or data. High-speed cables and devices are shielded; with ferrite beads recommended in series with all power and ground connector pins.

Attach Transient Surge

USB supports dynamic attach (i.e., "hot plug") of peripherals, as long as the initial load is less than the equivalent of a 44Ω resistor in parallel with 10µF. A current surge caused by a hot plug-in of a downstream device may momentarily reduce BPH voltage to other downstream devices below 4.00V. USB requires that each downstream port have a 120µF minimum bulk tantalum bypass capacitance close to the downstream connectors.[25] A hot-attach or downstream switch enable may cause a transient current surge due to capacitor charge current. Layout designed to reduce transient voltage droop may allow this transient current to be drawn through a protection device for a few microseconds. This surge current may cause electronic switches to briefly go into current-limiting, causing the error flag to fall. A simple 1ms RC filter in series with the error flag line to the hub controller prevents these transients from activating over-current shutdown at the hub, while allowing true over-current events to be reported.

---

[25] USB Specification, Rev. 1, ¶ 7.2.4.1

USB Power Management

Detach Transient Surge

When current in a wire is interrupted, the inductance of the wire causes a voltage spike as the magnetic field collapses. To reduce these spikes, which generate EMI, and to prevent damage to components, 0.01µF 25V ceramic bypass capacitors should be installed directly from Vbus pin to Ground pin at each port.[26] The peripheral upstream connector must have a 1µF minimum capacitor, (preferably at least 4.7µF, but less than 10µF equivalent including voltage regulator) peripheral bulk capacitance.[27]

8.2.10 Bus-Powered Hubs

Bus-powered hubs (BPH) obtain operating power from an upstream port, and may supply 100mA or more to each downstream port, if the input power budget is not exceeded.[28] BPH could be inside portable computers, keyboards or modems, or be autonomous. BPH may draw 100mA at startup from an upstream SPH port, increasing up to 500mA maximum after enumeration, apportioning the power to 100mA per downstream port and 100mA internally.

Keyboard

The most common BPH is expected to be a keyboard. A single cable will connect to a BPH near the user, probably in the monitor. A USB mouse, joystick, trackball or hand-scanner may be plugged into the keyboard.

Stand-alone

Many users will acquire USB peripherals gradually, so will need a stand-alone BPH to provide connectivity and power.

---

[26] USB Specification, Rev. 1, ¶ 7.2.4.2
[27] Intel USB Voltage Drop and Droop Measurement paper, Nov. 18, 1996. Part of Review Requests 94 and 112.(under revision, withdrawn from the USB web site)
[28] USB Specification, Rev. 1.1, ¶ 7.2.1 DWG Review Request 122, agreed 1/26/97

Chapter 8: Appendix

8.2.11 Bus-Powered Hub Requirements

USB specifies that Bus-Powered hubs (BPH) have power switching for their downstream ports, so that downstream current may be enumerated and controlled.[29] The upstream cable connector must receive a minimum voltage of 4.65V, while each downstream port provides a minimum of 4.40V at 100mA. A maximum of 100mA may be drawn from the upstream port when a BPH is first connected, equivalent to a 44Ω, 10µF load.[30]

Power Loads

BPH draw all of the power to any internal functions and to downstream ports from the upstream connector power pins. BPH may draw up to 100mA upon power-up. After configuration a total of 500mA is split between allocations to the hub, any embedded functions and the external ports. "Additionally, all external ports must be able to supply at least 100mA at any time, whether or not that port is currently connected and regardless of the power draw upon other ports in the same hub. For example, if a bus-powered hub has fewer than four external ports, it may be able to provide more than one load to any of its external ports as long as at least one load is available to each of the remaining ports. Any single external port may not be configured to draw more than one load unless the hub supports ganged overcurrent protection".[31]

Switched Power

Bus-powered hubs (BPH) are required to power- off all downstream ports until the hub is configured or receives a reset on its root port. Ports may be switched on or off under host software control. An implementation may provide power switching on a per port basis or have a single switch for all the ports (gang mode power control). Port reset requests do not affect the status of the

---

[29] USB Specification, Rev. 1, ¶ 11.7
[30] USB Specification, Rev. 1, ¶ 7.2.11
[31] USB Specification, Rev. 1.1, ¶ 7.2.1 DWG Review Request 122, agreed 1/26/97

USB Power Management

power switching for a port. A hub port must be powered on in order to perform connect detection from the downstream direction.[32]

Gang Mode

"The gang mode is a logic-OR mode; i.e. All ports in a gang will be switched on if any of the ports is switched on and all will need to be switched off for any to be switched off".[33] After enumeration by the host controller, up to 500mA may be drawn from the upstream port, 100mA for the internal hub controller and functions, and 100mA to each of up to four ports.

Suspend Mode Power

If downstream ports are in suspend mode, only 500µA per port may be drawn, including data line pull-up and pull-down resistors. For BPH, this current does not include current drawn by powered devices connected to downstream ports. When a BPH is in suspend, it must be able to provide the maximum current per downstream port (100mA).[34] Micrel's integrated power switches have <110µA per package operating current, with soft start to ramp capacitor charge inrush currents.

Reset

Reset is signaled downstream from a hub port. A BPH that receives a reset on its upstream port removes power from all downstream ports. All devices that receive the reset are set to their default address in the unconfigured state. Reset can wake a device from the suspended mode.[35]

---

[32] USB Specification, Rev. 1, ¶ 11.7
[33] USB Specification, Rev. 1.1, ¶ 11.7 DWG Review Request 104, agreed 10/10/96
[34] USB Specification, Rev. 1, ¶ 7.2.3
[35] USB Specification, Rev. 1, ¶ 7.1.4.3

Chapter 8: Appendix

Disabled

A downstream port must be powered-on in order to detect a connect event. When a port detects a connect event it changes to the disabled state. When in the disabled state the port output buffers are high-impedance, so the voltage level caused by pull-up resistors on the data lines may be assessed.[36]

Inrush Current

When power switches are turned-on, downstream capacitors may draw a high inrush current, causing a voltage droop upstream which could affect other circuits. USB requires that hot-plug or switched events draw no more current than a 10μF capacitor in parallel with a 44Ω load. High inrush current may also cause EMI, or damage tantalum capacitors. Micrel's power switch products provide soft start, limiting inrush current.

Capacitor Charge Current

When a voltage is applied to an uncharged capacitor, the average charge current is a function of capacitance times charging voltage over time. Tantalum capacitors may be damaged by high inrush currents caused by rapid onset of voltage.[37]

Resume Mode Power

When devices wake up they must limit the inrush current from upstream. The target maximum droop is 330mV. Devices must have a controlled power-up sequence preventing inrush current from exceeding BPH 500mA maximum.[38]

---

[36] USB Specification, Rev. 1, ¶ 11.2.3
[37] USB Specification, Rev. 1, ¶ 7.2.4.1
[38] USB Specification, Rev. 1, ¶ 7.2.3

8.2.12 Discrete Power Switch

Discrete P-channel MOSFET high-side switches are turned on by pulling their gate to ground with an external switch. For USB 5V applications, logic-level P-channel MOSFETs are generally double the price of similar on-resistance N-channel devices. Discrete N-channel high-side switches require an auxiliary power supply and logic-level converter to raise their gate voltage to 10-15V. Discrete switches require additional components for soft-start and to report over-current.

Discrete P-Channel Switch Surge

If P-channel MOSFET switch enable signals have a fast fall-time, the rapid charge current of output filter capacitors may cause an upstream power supply overcurrent flag to fall for 1-2ms. The hub controller could be programmed to ignore this inrush flag. An external resistor-capacitor network may be added to lengthen the fall-time, reducing the surge current.

8.2.13 Soft start

Micrel's MIC25xx power switches have a soft start of 1-2 ms, reducing the surge current - from the >10A initial current of a fast-start switch which turns on in 100μs - to a lower current ramping to 1A over 1 ms. N-channel MOSFET high-side switches, e.g., MIC2526, have a charge-pump which stretches the turn-on time to 1-2 mS.

Capacitor Inrush Control

USB requires that a device present an equivalent load maximum of 44Ω in parallel with 10μF when initially plugged in to a downstream port.

Chapter 8: Appendix

Hot Plug

If a high power device requires a capacitor larger than 10μF, a power switch inside the device could be used to limit inrush current.

Suspend Power

When a hub is suspended, downstream ports are still powered. The hub is allocated 500 μA plus 100 mA per downstream port. When downstream devices are suspended the maximum allowed average current per port is 500 μA, including the data line pull-up resistors and power switch operating current.[39]

Signal Termination

USB requires that cables be terminated at both ends. Hub downstream connector data pins have 15 kΩ resistors to ground. [40] A peripheral's upstream connector has 1.5 kΩ connected from a 3.3V +/-0.3V source to the D+ connector to indicate high speed, or to the D- connector for low speed operation.[41] The voltage source on the pull-up resistor should be derived or controlled from the power supplied from the upstream cable. Devices are prohibited from supplying current upstream on power or data lines.[42]

Dynamic Detach

When a device is quickly unplugged from the bus, with up to 500mA flowing in the cable, lead inductance may cause flyback voltage transients. Low capacitance, low inductance bypass capacitors (e.g., 0.01μF, 25V monolithic ceramic) should be installed close to each connector, from the power pin to the ground pin.[43] The

---

[39] USB Specification, Rev. 1, ¶ 7.1.4.4
[40] USB Specification, Rev. 1, ¶ 7.1.4.1
[41] USB Specification, Rev. 1, ¶ 7.1.3
[42] USB Specification, Rev. 1.1, ¶ 7.1.3 DWG Review Request 90, agreed 10/10/96
[43] USB Specification, Rev. 1, ¶ 7.2.4.2

downstream connector must have a 1μF capacitor, (preferably at least 4.7μF, but less than 10μF equivalent including voltage regulator) peripheral bulk capacitance.[44]

Voltage Drop

A 350mV maximum total drop is allowed from upstream Series A cable connector to BPH downstream socket connector. This drop includes the cable, connectors, switch and board. Standard length cables have a wire size chosen to produce a voltage drop of less than 190mV for two conductors at 500mA current (see Table 5). Shorter cables using the same gauge wire produce a smaller voltage drop, easing the requirement for power switch low on-resistance.

8.2.14 USB Cables

USB specifies the construction and wire gauge of cables used to connect host to hub to peripherals. USB specifies a maximum cable power wire resistance of 190mΩ, and a maximum cable length of 5m, (3m for a sub-channel cable,)[45] to minimize voltage drop and transmission time

Voltage Drop

USB allows maximum voltage drop of 350mV across a bus-powered hub, including the upstream cable, cable connectors, printed circuit board, control switch, and output connector. The drop is computed by multiplying the maximum current times the sum of all resistances in both the power cable and the ground return path. The largest portion of the voltage drop generally occurs in the two wires of an upstream BPH cable. A full four port current drain of four unit loads, plus 100mA for the USB controller, reduces the voltage

---

[44] Intel USB Voltage Drop and Droop Measurement paper, Nov. 18, 1996. Part of Review Requests 94 and 112.(under revision, withdrawn from the USB web site)
[45] USB Specification, Rev. 1, ¶ 6.1

from a minimum of 4.75V at the upstream end of the cable to 4.40V at each downstream BPH port.[46]

Cable Resistance

The resistance of copper wire is a function of its diameter, work hardening, annealing, and length. Table 3 (USB table 6-9) shows USB cable maximum length by wire size, based on 190mΩ maximum per power wire. Standard AWG20 cables have a lower resistance (179mΩ) since the length is limited to 5m to meet AC transmission timing constraints.

Table 3. Maximum Cable Length

Gauge	Resistance	Length (Maximum)
28	0.232Ω/m	0.81m
26	0.145Ω/m	1.31m
24	0.091Ω/m	2.08m
22	0.057Ω/m	3.33m
20	0.036Ω/m	5.0m*
		note:* AC signal limit

Connector Resistance

The standard USB mated connector is assumed to have a 30mΩ contact resistance per pin, including the solder connections to the PC board and cable wires, after 1,500 insertions. Since the voltage cannot be measured at the contact point, USB voltages are measured at the PC board side of a connector, from power pin to ground pin, not the power supply ground.

8.2.15 Detachable Cables

Detachable cables have different type connectors on each end to prevent false connections (e.g., downstream port of one hub to downstream port of a second hub).[47]

---

[46] USB Specification, Rev. 1, ¶ 6.4
[47] USB Specification, Rev. 1, table 6-9

USB Power Management

Standard Detachable Cables

A standard USB cable has a Series A connector on the upstream end, and a Series B connector on the downstream end, with AWG 28 data wires, 90Ω +/-15% impedance. Cables are not allowed to have the same series (A or B) connector on both ends, ensuring proper data and power flows.[48]

Bus-Powered Functions

A Bus-Powered function may draw up to 500mA from an SPH downstream port, using any standard USB cable. The maximum voltage drop across the cable will be less than 250mV under worst case conditions. No downstream ports are provided, so the bus-powered function may utilize 500mA.

8.2.16 Voltage Drop

There are two components of the voltage drop:

High Current in Upstream Cable

Use the 500 mA maximum current to compute the voltage drops in the upstream cable connector, cable wires and downstream cable header or connector.

The voltage drops are:

V_{con} SPH = 0.5A*30mΩ*2 pins = 30mV,

V_{con} BPH = 0.5A*30mΩ*2 pins = 30mV,

V_{wire} = 0.5A*R_{wire}*2*wire length (see table 6),

$V0.5$ = V_{con} SPH+ V_{con} BPH + V_{wire} = 60mV + V_{wire}.

---

[48] USB Specification, Rev. 1, ¶ 6.1

Unit Load Current in Downstream Cable

Use the 100mA maximum current to compute voltage drops in the PC board traces, power switch, output filter and downstream output connector.

The voltage drops are:

$Vpc = 0.1A*15m\Omega*2$ traces $= 3mV$,

$Vswitch = 0.4A*Ron, (0.14\Omega$ for MIC2525$) = 56mV$,

$Vfb = 0.1A*15m\Omega*2$ beads $= 3mV$,

$Vcon = 0.1A*30m\Omega*2$ pins $= 6mV$,

$V0.1 = Vpc + Vsw + Vfb + Vcon = 68mV$

Maximum Cable Length

Taking the sum of all of the voltage drops except the cable resistance from the 250mV maximum drop:

$Vwire = 250mV - 60mV = 190mV$.

From table 4, the maximum wire length for an AWG 20 wire pair is:

$Lwire = 190mV/0.5A/0.036\Omega/2m = 5.28m$, which is more than the maximum 5.0m allowed for cable length due to transmission time limits.

Low-Power devices

Low-Power devices are required to have an attached cable, less than 5m, 190mΩ maximum.

8.2.17 Layout

The power circuitry of USB printed circuit boards requires customized layout to minimize voltage drops, EMI and to maximize thermal dissipation.

Printed Circuit Board Traces

Printed circuit board traces for power conduction should be fabricated heavier than those used in normal digital layout practice, to reduce the in-circuit resistance in both the positive and ground traces. Each solder or header connection may be expected to contribute about 10 mΩ, emphasizing trace resistance reduction. Table 4 shows typical resistance in mΩ per inch for standard conductor widths and thickness.

Double-sided boards

Placing traces on both sides of the board, with no solder-mask on the solder-side, can cut the trace resistance in half. Solder plating on the solder-side also reduces trace resistance, by increasing trace thickness.

Ground Planes

Ground planes may have significant drops in through-hole vias. Short and wide same-surface traces are generally more effective in reducing voltage drops. Reduction of board, connector, and cable resistance losses is usually the most cost-effective solution to reducing voltage drops. Ground planes are also good thermal radiators and provide EMI suppression.

Table 4. Trace Resistance

Conductor Thickness	Conductor Width (in)	Resistance (mΩ / in)
½ oz/ ft²	0.025	39.3
	0.050	19.7
	0.100	9.8
	0.200	4.9
1 oz. /ft²	0.025	19.7
	0.050	9.8
	0.100	4.9
	0.200	2.5
2 oz. / ft²	0.025	9.8
	0.050	4.9
	0.100	2.5
	0.200	1.2
3 oz./ ft²	0.025	6.5
	0.050	3.2
	0.100	1.6

Chapter 8: Appendix

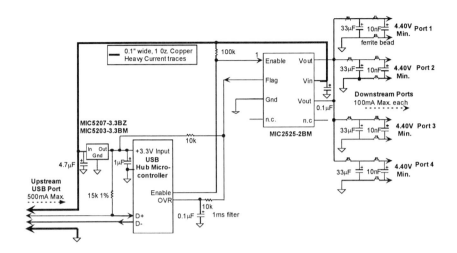

Figure 5. Ganged Output 4 Port Bus-Powered Hub

Figure 5 shows a single MIC2525 switch providing ganged power switching to four downstream ports. The single power switch controls all downstream ports, soft start limiting the inrush current to the downstream 33µF capacitors when the switch is enabled. The voltage drop at 400mA across the 140mΩ switch and other components is less than 100mV, providing a minimum of 4.40V at each 100mA loaded output port.

8.2.18 Micrel Power Switches

Micrel Semiconductor's logic-controlled high-side power switch integrated circuits are ideal for USB's power distribution requirements.

The MIC252x devices are high-side switches with built-in current limiting and thermal protection. Table 1 shows the main characteristics of the USB oriented switch products. All devices have logic inputs designed for either 3V or 5V signals, plus open-drain error flag outputs capable of operation to 7V in the 250x and

252x products. The switches are suitable for USB hosts, self-powered hubs and bus-powered hub.

These features provide all of the necessary functions to satisfy all of USB power distribution requirements.

Error Flag

Open-drain pull-down error flag outputs are available on MIC252x switches, indicating overcurrent or thermal shutdown.

Individual Switch Flag outputs

The single MIC25x5 and dual MIC 25x6 switches have an error flag associated with each output.

Logical "OR" Flag

Open-drain pull-down flag outputs may be connected to a single pull-up resistor, so that the output voltage is low if any of the flag outputs are low.

Soft start

A slow turn-on feature prevents high inrush currents when initially powering capacitive loads, as required by the USB specification.

Gate Drive Ramp

When the enable pin is activated, the drive voltage to the switch gate is gradually increased, causing a gradual decrease in on-resistance until the switch is in full conduction after 1 – 2ms.

Current limit

Built-in current limiting circuits protect the power supply and load.

Chapter 8: Appendix

Excessive Load

If the switch current exceeds the preset current limit, the output changes to a constant current mode.

Fold-Back Limiting

In order to protect the circuit from excessive power dissipation, current limiting is reduced as output voltage falls.

Open- Load Detection

MIC2505 and 2506 devices have an optional open-load feature, which utilizes an external 100kΩ pull-up on the load side. If a load is disconnected, the output is pulled high (above 1V) when Enable is off (the switch is off). The error flag is then pulled low to indicate an open load. Less than 25kΩ on the output indicates an attached load. Large capacitance on the switch output line, such as the required 120µF on downstream ports, may cause a momentary error flag as Enable is turned off. The required USB data line pull-down resistors provide sufficient load to deactivate this feature. MIC2505-1, 2505-2, and 252x devices do not have this feature.

Under Voltage Lockout

MIC25xx switches feature Undervoltage lockout (UVLO), which ensures that the output cannot be turned on if the power input is below 2.5V.

Samples and Evaluation Boards

Evaluation samples and boards for Micrel switches are available. Fill out a request form on Micrel's web page (http://www.micrel.com). A two-port board, with per-port switches (MIC2526), output filters and USB "A" connectors is available in surface-mount or through-hole (specify enable polarity: -1 logic high enable, -2 logic low enable). A four-port, ganged

MIC25x5 board is also available (specify MIC2505 or MIC2525 and enable polarity).

8.2.19 Summary

Micrel's logic-controlled high-side power switch integrated circuits are ideal for USB power switching applications. The MIC2505, 2506 and MIC252x power switches are suitable for USB hosts, self-powered hubs and bus-powered hubs. A slow turn-on feature prevents high inrush current when initially powering capacitive loads, as required by the USB specification. An open-drain flag output is available on the MIC25xx switches, indicating overcurrent or thermal shutdown to the USB hub controller.

8.3 Glossary

This section lists and defines terms and abbreviations used throughout this book. It is taken from the USB Specification Version 1.0, and is reprinted with permission.

Access.bus	The Access.bus is developed by the Access.bus Industry Group, based on the Philips I^{2}C technology and a DEC software model. Revision 2.2 specifies the bus for 100 kbs operation, but the technology has headroom to go up to 400 kbs.
ACK	Acknowledgment. Handshake packet indicating a positive acknowledgment.
Active Device	A device that is powered and not in the suspend state.
ADB	See Apple Desktop Bus.
APM	An acronym for Advanced Power Management. APM is a specification for managing suspend and resume operations to conserve power on a host system.
Apple Desktop Bus	An expansion bus used by personal computers manufactured by Apple Computer, Inc.
Asynchronous Data	Data transferred at irregular intervals with relaxed latency requirements.
Asynchronous Rate Adaptation	The incoming data rate, Fs_i, and the outgoing data rate, Fs_o, of the RA process are independent (i.e., no shared master clock).
Asynchronous Sample Rate Conversion	The incoming sample rate, Fs_i, and outgoing sample rate, Fs_o, of the SRC process are independent (i.e., no shared master clock).
Audio Device	A device that sources or sinks sampled analog data.
AWG#	The measurement of a wire's cross section as defined by the American Wire Gauge standard.
Babble	Unexpected bus activity that persists beyond a specified point in a frame.
Bandwidth	The amount of data transmitted per unit of time, typically bits per second (bps) or bytes per second (Bps).

Glossary

Big Endian	A method of storing data that places the most significant byte of multiple byte values at a lower storage addresses. For example, a word stored in big endian format places the least significant byte at the higher address and the most significant byte at the lower address. See Little Endian.
Bit	A unit of information used by digital computers. Represents the smallest piece of addressable memory within a computer. A bit expresses the choice between two possibilities and is typically represented by a logical one (1) or zero (0).
Bit Stuffing	Insertion of a "0" bit into a data stream to cause an electrical transition on the data wires allowing a PLL to remain locked.
bps	Transmission rate expressed in bits per second.
Bps	Transmission rate expressed in bytes per second.
Buffer	Storage used to compensate for a difference in data rates or time of occurrence of events, when transmitting data from one device to another.
Bulk Transfer	Non-periodic, large, bursty communication typically used for a transfer that can use any available bandwidth and also be delayed until bandwidth is available.
Bus Enumeration	Detecting and identifying Universal Serial Bus devices.
Byte	A data element that is eight bits in size.
Capabilities	Those attributes of a Universal Serial Bus device that can be administered by the host.
Characteristics	Those qualities of a Universal Serial Bus device that are unchangeable; for example, the device class is a device characteristic.
CHI	An acronym for Concentration Highway Interface. CHI is a full duplex time division multiplexed serial interface for digitized voice transfers in communications systems. The current specification supports data transfer rates up to 4.096 Mbs.
Client	Software resident on the host that interacts with host software to arrange data transfer between a function and the host. The client is often the data provider and consumer for transferred data.
COM Port	Communications port. On personal computers, an eight-bit asynchronous serial port is typically used.

Configuring Software	The host software responsible for configuring a Universal Serial Bus device. This may be a system configurator or software specific to the device.
Control Pipe	Same as a message pipe.
Control Transfer	One of four Universal Serial Bus Transfer Types. Control transfers support configuration/command/status type communications between client and function.
CRC	See Cyclic Redundancy Check.
CTI	Computer Telephony Integration.
Cyclic Redundancy Check	A check performed on data to see if an error has occurred in transmitting, reading, or writing the data. The result of a CRC is typically stored or transmitted with the checked data. The stored or transmitted result is compared to a CRC calculated for the data to determine if an error has occurred.
Default Address	An address defined by the Universal Serial Bus Specification and used by a Universal Serial Bus device when it is first powered or reset. The default address is 00h.
Default Pipe	The message pipe created by Universal Serial Bus system software to pass control and status information between the host and a Universal Serial Bus device's Endpoint 0.
Device	A logical or physical entity that performs a function. The actual entity described depends on the context of the reference. At the lowest level, device may refer to a single hardware component, as in a memory device. At a higher level, it may refer to a collection of hardware components that perform a particular function, such as a Universal Serial Bus interface device. At an even higher level, device may refer to the function performed by an entity attached to the Universal Serial Bus; for example, a data/FAX modem device. Devices may be physical, electrical, addressable, and logical. When used as a non-specific reference, a Universal Serial Bus device is either a hub or a function.

Glossary

Device Address	The address of a device on the Universal Serial Bus. The Device Address is the Default Address when the Universal Serial Bus device is first powered or reset. Hubs and functions are assigned a unique Device Address by Universal Serial Bus software.
Device Endpoint	A uniquely identifiable portion of a Universal Serial Bus device that is the source or sink of information in a communication flow between the host and device.
Device Resources	Resources provided by Universal Serial Bus devices, such as buffer space and endpoints. See Host Resources and Universal Serial Bus Resources.
Device Software	Software that is responsible for using a Universal Serial Bus device. This software may or may not also be responsible for configuring the device for use.
DMI	An acronym for Desktop Management Interface. A method for managing host system components developed by the Desktop Management Task Force.
Downstream	The direction of data flow from the host or away from the host. A downstream port is the port on a hub electrically farthest from the host that generates downstream data traffic from the hub. Downstream ports receive upstream data traffic.
Driver	When referring to hardware, an I/O pad that drives an external load. When referring to software, a program responsible for interfacing to a hardware device; that is, a device driver.
DWORD	Double word. A data element that is 2 words, 4 bytes, or 32 bits in size.
Dynamic Insertion and Removal	The ability to attach and remove devices while the host is in operation.
E^2PROM	See EEPROM.
EEPROM	Electrically Erasable Programmable Read Only Memory. Non-volatile re-writeable memory storage technology.
End User	The user of a host.
Endless Loop	See Loop, Endless.

Endpoint	See Device Endpoint.
Endpoint Address	The combination of a Device Address and an Endpoint Number on a Universal Serial Bus device.
Endpoint Number	A unique pipe endpoint on a Universal Serial Bus device.
EOF1	End of frame timing point #1. Used by the hub to monitor and disconnect bus activity persisting near or past the end of a frame.
EOF2	End of frame timing point #2. Used by hubs to detect bus activity near the end of frame.
EOP	End of packet.
Fs	See Sample Rate.
False EOP	A spurious, usually noise-induced, event that is interpreted by a packet receiver as an end of packet.
FireWire	Apple Computer's implementation of the IEEE 1394 bus standard.
Frame	The time from the start of one SOF token to the start of the subsequent SOF token. A frame consists of a series of transactions.
Frame Pattern	A sequence of frames that exhibits a repeating pattern in the number of samples transmitted per frame. For a 44.1 kHz audio transfer, the frame pattern could be nine frames containing 44 samples followed by one frame containing 45 samples.
Full-duplex	Computer data transmission occurring in both directions simultaneously.
Function	A Universal Serial Bus device that provides a capability to the host. For example, an ISDN connection, a digital microphone, or speakers.
GeoPort	A serial bus developed by Apple Computer, Inc. Current specification of the GeoPort supports data transfer rates up to 2 Mbs and provides point to point connectivity over a radius of 4 ft.
Handshake Packet	A packet that acknowledges or rejects a specific condition. For examples, see ACK and NACK.
Host	The host computer system where the Universal Serial Bus host controller is installed. This includes the host hardware platform (CPU, bus, etc.) and the operating system in use.
Host Controller	The host's Universal Serial Bus interface.

Glossary

Host Controller Driver	The Universal Serial Bus software layer that abstracts the host controller hardware. Host Controller Driver provides an SPI for interaction with a host controller. Host Controller Driver hides the specifics of the host controller hardware implementation.
Host Resources	Resources provided by the host, such as buffer space and interrupts. See Device Resources and Universal Serial Bus Resources.
Hub	A Universal Serial Bus device that provides additional connections to the Universal Serial Bus.
Hub Tier	The level of connect within a USB network topology given as the number of hubs that that the data has to flow through.
I^2C	Acronym for the Inter-Integrated Circuits serial interface. The I^2C interface was invented by Philips Semiconductors.
IEEE 1394	A high performance serial bus. The 1394 is targeted at hard disk and video peripherals, which may require bus bandwidth in excess of 100 Mb/s. The bus protocol supports both isochronous and asynchronous transfers over the same set of four signal wires. See also FireWire.
Industry Standard Architecture	The 8 and/or 16-bit expansion bus (ISA) for IBM AT or XT compatible computers.
Integrated Services Data Network	An internationally accepted standard for voice, data, and signaling using public, switched telephone networks. All transmissions are digital from end-to-end. Includes a standard for out-of-band signaling and delivers significantly higher bandwidth than POTS.
Interrupt Request	A hardware signal that allows a device to request attention from a host. The host typically invokes an interrupt service routine to handle the condition which caused the request.
Interrupt Transfer	One of four Universal Serial Bus Transfer Types. Interrupt transfer characteristics are small data, non periodic, low frequency, bounded latency, device initiated communication typically used to notify the host of device service needs.
IRP	Interrupt Request Packet

IRQ	See Interrupt Request.
ISA	See Industry Standard Architecture.
ISDN	See Integrated Services Data Network.
Isochronous Data	A stream of data whose timing is implied by its delivery rate.
Isochronous Device	An entity with isochronous endpoints, as defined in the USB specification, that sources or sinks sampled analog streams or synchronous data streams.
Isochronous Sink Endpoint	An endpoint that is capable of consuming an isochronous data stream.
Isochronous Source Endpoint	An endpoint that is capable of producing an isochronous data stream.
Isochronous Transfer	One of four Universal Serial Bus Transfer Types. Isochronous transfers are used when working with isochronous data. Isochronous transfers provide periodic, continuous communication between host and device.
Jitter	A tendency toward lack of synchronization caused by mechanical or electrical changes. More specifically, the phase shift of digital pulses over a transmission medium.
kbs	Transmission rate expressed in kilobits per second.
kBs	Transmission rate expressed in kilobytes per second.
Line Printer Port	A port used to access a printer. On most personal computers, an eight-bit parallel interface is typically used.
Little Endian	Method of storing data that places the least significant byte of multiple byte values at lower storage addresses. For example, a word stored in little endian format places the least significant byte at the lower address and the most significant byte at the next address. See Big Endian.
LOA	Loss of bus activity characterized by a start of packet without a corresponding end of packet.
Loop, Endless	See Endless Loop.
LPT Port	See Line Printer Port.
LSB	Least Significant Bit.
Mbs	Transmission rate expressed in megabits per second.

Glossary

MBs	Transmission rate expressed in megabytes per second.
MCA	See Micro Channel Architecture.
Message Pipe	A pipe that transfers data using a request/data/status paradigm. The data has an imposed structure which allows requests to be reliably identified and communicated.
Micro Channel Architecture	A 32-bit expansion bus (MCA) used on some IBM PS/2 compatible computers.
Modem	An acronym for Modulator/Demodulator. Component that converts signals between analog and digital. Typically used to send digital information from a computer over a telephone network, which is usually analog.
MSB	Most Significant Bit.
NACK	Negative Acknowledgment. Handshake packet indicating a negative acknowledgment.
Non Return to Zero Invert	A method of encoding serial data in which ones and zeroes are represented by opposite and alternating high and low voltages where there is no return to zero (reference) voltage between encoded bits. Eliminates the need for clock pulses.
NRZI	See Non Return to Zero Invert.
Object	Host software or data structure representing a Universal Serial Bus entity.
OHCI	Open Host Controller Interface is a method used by the USB host PC to communicate with other devices. Originally defined by Microsoft.
Packet	A bundle of data organized in a group for transmission. Packets typically contain three elements: control information (e.g., source, destination, and length), the data to be transferred, and error detection and correction bits.
Packet Buffer	The logical buffer used by a Universal Serial Bus device for sending or receiving a single packet. This determines the maximum packet size the device can send or receive.
Packet ID	A field in a Universal Serial Bus packet that indicates the type of packet, and by inference the format of the packet and the type of error detection applied to the packet.
PBX	See Private Branch eXchange.

PCI	See Peripheral Component Interconnect.
PCMCIA	See Personal Computer Memory Card Industry Association.
Peripheral Component Interconnect	A 32- or 64-bit, processor independent, expansion bus used on personal computers.
Personal Computer Memory Card International Association	The organization that standardizes and promotes PC Card technology.
Phase	A token, data, or handshake packet; a transaction has three phases.
Physical Device	A device that has a physical implementation; e.g. speakers, microphones, and CD players.
PID	See Packet ID.
Pipe	A logical abstraction representing the association between an endpoint on a device and software on the host. A pipe has several attributes; for example, a pipe may transfer data as streams (Stream Pipe) or messages (Message Pipe).
Plain Old Telephone Service	Basic service supplying standard single line telephones, telephone lines, and access to public switched networks (POTS).
Plug and Play	A technology for configuring I/O devices to use non-conflicting resources in a host. Resources managed by Plug and Play include I/O address ranges, memory address ranges, IRQs, and DMA channels.
PnP	See Plug and Play.
Polling	Asking multiple devices, one at a time, if they have any data to transmit.
POR	See Power On Reset.
Port	Point of access to or from a system or circuit. For Universal Serial Bus, the point where a Universal Serial Bus device is attached.
POTS	See Plain Old Telephone Service.
Power On Reset	Restoring a storage device, register, or memory to a predetermined state when power is applied.
PLL	Phase Locked Loop. A circuit that acts as a phase detector to keep an oscillator in phase with an incoming frequency.

Glossary

Private Branch eXchange	A privately owned telephone switching system, which is not regulated as part of the public telephone network.
Programmable Data Rate	Either a fixed data rate (single frequency endpoints), a limited number of data rates (32 kHz, 44.1 kHz, 48 kHz, ...), or a continuously programmable data rate. The exact programming capabilities of an endpoint must be reported in the appropriate class-specific endpoint descriptors.
Protocol	A specific set of rules, procedures, or conventions relating to format and timing of data transmission between two devices.
RA	See Rate Adaptation.
Rate Adaptation	The process by which an incoming data stream, sampled at Fs_i is converted to an outgoing data stream, sampled at Fs_o with a certain loss of quality, determined by the rate adaptation algorithm. Error control mechanisms are required for the process. Fs_i and Fs_o can be different and asynchronous. Fs_i is the input data rate of the RA; Fs_o is the output data rate of the RA.
Request	A request made to a Universal Serial Bus device contained within the data portion of a SETUP packet.
Retire	The action of completing service for a transfer and notifying the appropriate software client of the completion.
Root Hub	A Universal Serial Bus hub directly attached to the host controller. This hub is attached to the host; tier 0.
Root Port	The upstream port on a hub.
Sample	The smallest unit of data on which an endpoint operates; a property of an endpoint.
Sample Rate (Fs)	The number of samples per second, expressed in Hertz.
Sample Rate Conversion	A dedicated implementation of the RA process for use on sampled analog data streams. The error control mechanism is replaced by interpolating techniques.
SCSI	See Small Computer Systems Interface.
Service	A procedure provided by an SPI (System Programming Interface).

Chapter 8: Appendix

Service Interval	The period between consecutive requests to a Universal Serial Bus endpoint to send or receive data.
Service Jitter	The deviation of service delivery from its scheduled delivery time.
Service Rate	The number of services to a given endpoint per unit time.
Small Computer Systems Interface	A local I/O bus that allows peripherals to be attached to a host using generic system hardware and software.
SOF	An acronym for Start of Frame. The SOF is the first transaction in each frame. SOF allows endpoints to identify the start of frame and synchronize internal endpoint clocks to the host.
SPI	See System Programming Interface.
SRC	See Sample Rate Conversion.
Stage	One part of the sequence composing a control transfer; i.e., the setup stage, the data stage, and the status stage.
Stream Pipe	A pipe that transfers data as a stream of samples with no defined Universal Serial Bus structure.
Synchronization Type	A classification that characterizes an isochronous endpoint's capability to connect to other isochronous endpoints.
Synchronous RA	The incoming data rate, Fs_i, and the outgoing data rate, Fs_o, of the RA process are derived from the same master clock. There is a fixed relation between Fs_i and Fs_o.
Synchronous SRC	The incoming sample rate, Fs_i, and outgoing sample rate, Fs_o, of the SRC process are derived from the same master clock. There is a fixed relation between Fs_i and Fs_o.
System Programming Interface	A defined interface to services provided by system software.
TDM	See Time Division Multiplexing.
Termination	Passive components attached at the end of cables to prevent signals from being reflected or echoed.

Glossary

Time Division Multiplexing	A method of transmitting multiple signals (data, voice, and/or video) simultaneously over one communications medium by interleaving a piece of each signal one after another.
Time-out	The detection of a lack of bus activity for some predetermined interval.
Token Generator	See Initiator.
Token Packet	A type of packet that identifies what transaction is to be performed on the bus.
Transaction	The delivery of service to an endpoint; consists of a token packet, optional data packet, and optional handshake packet. Specific packets are allowed/required based on the transaction type.
Transfer	One or more bus transactions to move information between a software client and its function.
Transfer Type	Determines the characteristics of the data flow between a software client and its function. Four Transfer types are defined: control, interrupt, bulk, and isochronous.
Turnaround Time	The time a device needs to wait to begin transmitting a packet after a packet has been received to prevent collisions on Universal Serial Bus. This time is based on the length and propagation delay characteristics of the cable and the location of the transmitting device in relation to other devices on Universal Serial Bus.
UHCI	Universal Host Controller Interface is a method the host PC uses to communicate with various devices. Originally defined by Intel.
Universal Serial Bus	A collection of Universal Serial Bus devices and the software and hardware that allow them to connect the capabilities provided by functions to the host.
Universal Serial Bus Device	Includes hubs and functions. See device.
Universal Serial Bus Interface	The hardware interface between the Universal Serial Bus cable and a Universal Serial Bus device. This includes the protocol engine required for all Universal Serial Bus devices to be able to receive and send packets.

Universal Serial Bus Resources	Resources provided by Universal Serial Bus, such as bandwidth and power. See Device Resources and Host Resources.
Universal Serial Bus Software	The host-based software responsible for managing the interactions between the host and the attached Universal Serial Bus devices.
USB	See Universal Serial Bus.
USBD	See Universal Serial Bus Driver.
Universal Serial Bus Driver	The host resident software entity responsible for providing common services to clients that are manipulating one or more functions on one or more Host Controllers.
Upstream	The direction of data flow towards the host. An upstream port is the port on a device electrically closest to the host that generates upstream data traffic from the hub. Upstream ports receive downstream data traffic.
Virtual Device	A device that is represented by a software interface layer; e.g., a hard disk with its associated device driver and client software that makes it able to reproduce an audio .WAV file.
WFEOF2	Wait for EOF2 point. One of the four hub repeater states.
WFEOP	Wait for end of packet. One of the four hub repeater states.
WFSOF	Wait for start of frame. One of the four hub repeater states.
WFSOP	Wait for start of packet. One of the four possible hub repeater states.
Word	A data element that is two bytes or 16 bits in size.

8.4 Frequently Asked Questions (FAQ)

This section consists of the FAQ portion of the USB Implementer's Forum web site. This site is a service of the USB Implementers Forum staff. The primary charter for that staff is to support the USB Implementers Forum member companies with Developers Conferences, Spec Technical Support, and Compatibility Workshops. USB-IF also provides marketing exposure for the members as a unified group of companies producing USB products at trade shows such as Comdex and PC Expo.

The secondary charter is to provide the industry at large with any information that is available to foster understanding and development of USB. This FAQ is a tool in that information distribution. The author of this book gratefully acknowledges the permission from the USB-IF to reprint this information. Some of the content has been updated by the author.

1. What is USB?
2. What kind of peripherals will USB allow me to hook up to my PC?
3. How does it work?
4. Will I need to purchase special software to run USB peripherals?
5. Will USB peripherals cost more?
6. Is there a Mac version of the standard?
7. Are there USB products out right now?
8. How can USB be used between two hosts, like a laptop and a desktop?
9. How does USB compare with Sony's FireWire/IEEE 1394 standard?
10. When it is available, will FireWire replace USB?
11. Who created USB anyway?

12. What are the Intellectual Property issues with USB, is there a license, what does it cost, what is the "Reciprocal Covenent Agreement" I've heard about?
13. What is the USB-IF?
14. What are the benefits of USB-IF?
15. How do I join USB-IF?
16. How do I get in touch with USB-IF?
17. What about this OHCI and UHCI?
18. Where do I get the SIE VHDL?
19. Where can I get a spec?
20. Is there a newsgroup for USB?
21. How do I get a USB vendor ID?
22. Where can we get EMC testing peripherals?
23. Hasn't someone informed the trade press that there is a need for both Firewire and USB?
24. Is the USB bus going to have a long distance 50-200 meter extension (possibly fiber) for these large customers that need the capability?
25. Will legacy device support be in the formal USB spec? When?
26. Will source code for driving HCI chips be made available?
27. When a device is detached, its device driver is unloaded If that device is re-inserted, would its driver be reloaded?
28. Are there any plans to increase the bus bandwidth of USB in the future to 2x, 3x?
29. Can someone clarify the difference and applications for series "A" and series "B" connectors?
30. What is the difference between a root hub and normal hub in terms of hardware and software?

Frequently Asked Questions (FAQ)

    31. Is USB a viable Bus for peripherals like CD-R, tape, or hard disk drives?

    32. The programming spec for UHCI is not publicly available. When can one get the UHCI spec?

    33. How do I get a USB PDK system?

8.4.1 What is USB?

USB is a peripheral bus standard developed by PC and telecom industry leaders -- Compaq, DEC, IBM, Intel, Microsoft, NEC, and Northern Telecom -- that will bring Plug and Play to computer peripherals outside the box, eliminating the need to install cards into dedicated computer slots and reconfigure the system. Personal computers equipped with USB will allow computer peripherals to be automatically configured as soon as they are physically attached - without the need to reboot or run setup. USB will also allow multiple devices -- up to 127 -- to run simultaneously on a computer, with peripherals such as monitors and keyboards acting as additional plug-in sites, or hubs.

8.4.2 What kind of peripherals will USB allow me to hook up to my PC?

You name it: monitor controls, audio I/O devices, telephones, modems, keyboards, mice, 4x and 6x CD ROM drives, joysticks, tape and floppy drives, and imaging devices such as scanners and printers. USB's 12 megabit/second data rate will also accommodate a whole new generation of peripherals, including MPEG-2 video-based products, data gloves, and digitizers. Also, since computer-telephony integration is expected to be a big growth area for PCs, USB will provide a low-cost interface for Integrated Services Digital Network (ISDN) and digital PBXs

8.4.3 How does it work?

Drawing its intelligence from the host PC, USB will detect when devices are added and removed. The bus automatically determines what host resource, including driver software and bus bandwidth, is needed by each peripheral, and makes those resources available without user intervention. Users with a USB-equipped PC will be able to switch out compatible peripherals as needed as easily as plugging in a lamp.

8.4.4 Will I need to purchase special software to run USB peripherals?

The Windows operating system (since OSR 2.1 release on October 29, 1996) comes already equipped with the feature (called "drivers") that allows your PC to recognize USB peripherals. Ultimately, you will not need to purchase or install additional software for each new peripheral. However, the availability of new peripheral products (including those never-before-seen) may create a gap between the peripheral availability and operating system upgrades. In these cases you will receive a diskette with your new USB peripheral containing the necessary drivers.

8.4.5 Will USB peripherals cost more?

USB peripherals will be competitively priced with the peripherals available on the market today.

8.4.6 Is there a Mac version of the standard?

Apple is working on products with USB interfaces.

8.4.7 Are there USB products out right now?

Yes. Almost all new PC designs from major vendors shipping today have USB connections on the motherboard and the correct Windows OS to make them work. There are also many products used to design and build USB systems, such as connectors, chipsets, and board-level computers. USB peripherals, including keyboards, monitors, mice, and joysticks are starting to appear in 1997.

Frequently Asked Questions (FAQ)

8.4.8 How can USB be used between two hosts, like a laptop and a desktop?

The answer is a small adapter that would appear as a device to each USB system desiring connections. Two USB peripheral microcontrollers sharing a buffer memory would be a quick solution and could sell for under $50. The packaging could be as streamlined as a small blob (dongle) in the middle of the cable or maybe even a slightly large connector shell at one end and nothing in the middle of the cable. A cable like this could also perform hub functions for very little extra cost to produce a higher-value product.

8.4.9 How does USB compare with the FireWire/IEEE 1394 standard?

They differ most in terms of application focus, availability, and price. The USB feature is available now and will address more traditional PC connections, like keyboards, mice, joysticks, and handheld scanners. However, USB's data rate (12 Mb/s) is more than adequate for many consumer applications including more-advanced computer game devices, high-fidelity audio, and highly-compressed video, like MPEG-1 and MPEG-2. Most importantly, the USB feature will add nothing to system cost.

FireWire will only be available in low volume until late 1997. FireWire will target high-bandwidth consumer electronics connections to the PC -- like digital camcorders, cameras, and digital video disc players.

8.4.10 When it is available, will FireWire replace USB?

No. The two technologies target different peripheral connections and will therefore be complementary. When FireWire becomes more prevalent, in about two years, it will be up to individual consumers what features they want on their new PCs. It seems likely that, in the future, PCs will have both USB and FireWire connection ports.

8.4.11 Who created USB, anyway?

USB was developed by a group of seven companies that saw a need for an interconnect to enable the growth of the blossoming Computer Telephony Integration Industry. The seven promoters of the USB definition are; Compaq, Digital Equipment Corp, IBM PC Co., Intel, Microsoft, NEC, and Northern Telecom.

8.4.12 What are the intellectual property issues with USB? Is there a license, what does it cost, and what is the "Reciprocal Covenent Agreement" I've heard about?

USB is royalty free because the promoters that created the specification agreed to allow anyone to build products around the spec without any charge. The promoters have signed an intellectual property (IP) agreement, which promises there will be no lawsuits based on any IP incorporated within the specification. The Reciprocal Covenant Agreement is a copy of that agreement with a place for any implementer to sign and return to USB-IF administration in order to go on record as having seen and understood the agreement. The Reciprocal Covenant is made available to anyone (USB-IF member or not) to clarify the USB license agreement.

8.4.13 What is the USB-IF?

USB Implementers Forum is a support organization formed by the seven promoters of USB to help speed development of high quality compatible devices using USB. The organization is administered by personnel supplied by the promoter companies and is funded by membership fees.

The organization is non-profit and uses all funds for promotion of USB products or technical education of implementers of USB products. Most of the activities USB Implementers Forum sponsors are organized to be breakeven events. Large presence shows such as Comdex and the USB Compliance Workshops (PlugFests) consume most of the organization's budget.

8.4.14 What are the benefits of USB-IF?

We are constantly looking for more ways to encourage and accelerate USB product development. The following list is a collection of benefits that USB-IF has committed to its members.

    Free vendor ID

    Free Technical support

    White papers, design guides, app notes etc.

    Discounted attendance at Developer Conferences

    Participation in Compatibility Workshops

    Invitation to participate in Marketing Events

    Company listing in USB key contacts list

    Participation in USB committed products list

    Participation in industry discussion mail list

    Five free copies of each hardcopy spec. An electronic version of the spec is available at www.usb.org. Hard copies are available from Annabooks.

    USB SIE VHDL at no charge

8.4.15 How do I join USB-IF?

A membership application can be downloaded from the USB-IF web site at http://www.usb.org. Instructions for processing are on the application.

8.4.16 How do I get in touch with USB-IF?

    The home page (http://www.usb.org) has contact buttons for all of the common types of questions.

    Administrative questions on USB-IF or events call 503-264-0590 or FAX 503-693-7975.

    Mail questions and applications to; USB-IF, 2111 NE 25th. Ave. MS JF2-51, Hillsboro OR 97124.

8.4.17 What about OHCI and UHCI?

Both OHCI, and UHCI are USB spec compatible and provide an interface to different hardware host controller implementations. Multiple implementations of hardware host controllers allow for evolution and creativity within the USB spec. Please also see the question in section 8.4.33.

8.4.18 Where do I get the SIE VHDL?

Intel gave the VHDL source for its Serial Interface Engine to USB-IF for distribution to its members. All members with valid email addresses receive the SIE VHDL by email. If you have not gotten it and you are a member of USB-IF check with administration to make sure that your email address is correct. The full source code for the SIE VHDL in a peripheral constitutes the SIE VHDL. A peripheral is composed of other logic besides the SIE and that logic of course is unique to each peripheral and is not available in the SIE VHDL. At this time the latest version of VHDL is .42 and there is no plan for an update in progress. The SIE VHDL is not available from USB-IF for non-members.

8.4.19 Where can I get a spec?

The current spec is available for download from the home page (http://www.usb.org). Hard copies are available from Annabooks.

8.4.20 Is there a newsgroup for USB?

There is an email function called the Reflector set up for USB-IF members that allows discussion and interaction between companies. There is no monitoring or censorship except for the guideline to use the list only for USB-related communications. This is not a real news group since it operates through email and it does not preserve history for all to see. If you save answers to questions that appear on it we would appreciate your re-sending them as needed. For current product directory and availability see www.usbnews.com.

8.4.21 How do I get a USB vendor ID?

USB Implementers Forum members are issued a vendor ID free when they join. Non-members can obtain a vendor ID by contacting the USB-IF administration number listed in the answer to the contact question. There is a $200 handling fee for non-members.

8.4.22 Where can we get EMC testing peripherals?

In order to get EMC compliance it is necessary for all external connectors of the equipment to be connected to a peripheral device and that device must be exercised during testing. Both low-speed and high-speed testing is also required. Any of the products of USB-IF members that have shown prototypes at our developers conferences or trade shows would be suitable. The evaluation kit that comes with the Intel PDK would be suitable and is also available separately. Even a passive resistive load at the end of a normal length of cable would be suitable if software to exercise the cable was used.

8.4.23 Hasn't someone informed the trade press that there is a need for Firewire and USB?

Yes, the press has always been known for their auditory acuity.

8.4.24 Is the USB bus going to have a long distance 50-200 meter extension (possibly fiber) for these large customers that need the capability?

USB is intended for desktop (or laptop) peripheral interface, and 200 meters seems like a rather large desktop. Still, many member companies have talked about longer distance applications and are thinking of creating the products needed to accomplish them. One possibility is an extension device that looks like a hub to the USB bus from both sides, but utilizes another protocol (such as fiber) between the endpoints of the cable. Each end would translate USB electrical signaling to or from a long distance signaling. While this is possible, there are issues regarding packet protocol and latency that must be considered to maintain USB compatibility.

8.4.25 Will legacy device support be in the formal USB spec? When?

Legacy support is not a USB spec issue. There is a class document being developed that addresses the legacy questions. It is available now in a 0.9 revision from our class document download section.

8.4.26 Will source code for driving HCI chips be made available?

The HCI drivers are supplied as a part of the OS software stack. Obtaining source code for those drivers would have to be discussed with the OS vendors. The manufacturers of host controllers may have some test drivers or production drivers, but again, access to the source code would have to be discussed with the owners of the code.

8.4.27 When a device is detached, its device driver is unloaded. If that device is re-inserted, would its driver be reloaded?

Yes, dynamic configuration and initialization by the OS includes automatically loading and unloading the drivers as needed.

8.4.28 Are there any plans to increase the bus bandwidth of USB in the future to 2x, 3x?

No, USB was designed for a desktop peripheral interface and has a performance/cost point for today's peripherals. A new interface, such as 1394, for future high-speed peripherals may develop.

8.4.29 Can someone clarify the difference and applications for series A and series B connectors?

The series A connector is intended for all USB devices. It has a plug for a peripheral and a socket for a PC platform. In most cases a USB cable should be captive (molded in) to its peripheral. This saves connector cost, eliminates incompatibilities due to power drop in a cable, and simplifies the user connection task. There are some cases where a captive cable is prohibitive. A very large heavy device may not be able to tolerate dangling cables that cannot be removed and some devices that are only occasionally connected, but have a useful function when not connected are good examples. The series B connector was created for such applications. The two connectors are different to prevent connections that violate the USB architecture topology.

8.4.30 What is the difference between a root hub and normal hub in terms of hardware and software?

All hubs are identical from a software viewpoint (notwithstanding the bus-powered and self-powered differences). A root hub is simply the first hub encountered during enumeration. In many implementations the root hub can be integrated into the same silicon as the host controller to save cost.

8.4.31 Is USB a viable bus for peripherals like CD-R, tape, or hard disk drives?

The viability depends on the definition of acceptable performance point. If any of these devices are for frequent use then I would want a permanent installation both for performance and mechanical integration. USB is not intended to be an inside-the-box permanent connection for high-speed peripherals. If the use is occasional or is for a peripheral that is shared between many computers, I would think that USB performance would be more than sufficient. The convenience of USB and the ubiquitous connection that USB will bring would outweigh blazing transfer rates. Still, USB will provide CD transfer rates up to 4x or 6x drives (not enough for re-writeable drives) and better transfer rates than the typical LPT CONNECTED tape drive, floppy drive or removable hard disk.

8.4.32 The programming spec for UHCI is not publicly available. When can one get the UHCI spec?

Actually the UHCI spec is publicly available on the Intel web site. The USB web site has a link to both the OHCI and UHCI public information sites.

8.4.33 How do I get a USB PDK system?

The PDK was a standard Intel product, and all Intel sales representatives and distributors should be able to give you details of price and availability. The need for that product is vanishing now that USB systems are available in the retail market. Intel recently changed the contents of the PDK as described in this letter to current PDK owners: (other silicoon developers such as Cypress and USAR also sell developer's kits. Contact information is available at www.usbnews.com.)

> Intel Corporation 5200 N.E. Elam Young Parkway Hillsboro, OR 97124-6497 1-800-628-8686 (Option 1,3,1) Universal Serial Bus Peripheral Developers Kit

September, 1996

Dear Universal Serial Bus Developer,

Intel's USB Peripheral Developer Kit containing the USB Host PC and 930 Evaluation Kit has been a huge success. These kits have been deployed widely across the industry and in all geographies. Complemented by the USB Implementer's Forum compliance workshops and Microsoft's beta program, the industry is well on its way to delivering end-user products. OEMs are already shipping PC host systems with Intel's PCI set silicon and USB connectors.

The Peripheral Developers Kit hardware and software is changing. The 930 Evaluation Kit will continue to be available from Intel (order code USB930EVALKIT). This kit has been expanded to include the Single Step Transaction Debugger (SSTD). Intel will continue to upgrade the evaluation kit, as it is the most popular platform for USB peripheral developers. Updates to the SSTD will be available from Intel's Bulletin Board System at 916-356-3600 or the World Wide Web. Contact your local Intel sales representative for information on how to register for these updates. The Microsoft's WDM beta, which has equivalent functionality to the Windows VxD-based stack that shipped with your systems, is now available from Microsoft. We strongly suggest you to enroll immediately in their beta program in order to stay up-to-date with the releases. Microsoft expects you to sign and return the appropriate NDA to allow them to send you the current beta release. Please follow up directly with Microsoft on the highest urgency. The primary reason for this transition is to focus your device and device driver development on the final production target.

To sign up for Microsoft's beta program, please send a request to usbbeta@microsoft.com and make sure to add the word "spatula" in the subject header to get on their priority list. Also, please include the following info in your message: First Name, Last Name, Title, Company Name, Address, City, State, Zip, Country, Phone#, Fax#, and internet address.

Some of the initial USB host PCs were manufactured with an early stepping (A1) of the PCI chipset. While the Intel VxD stack works with these systems, the Microsoft Beta release works only with systems built using the later stepping (B0). Intel will be sending a replacement motherboard at no charge to purchasers of these early systems. You should expect to receive the upgrade hardware in October. Please do NOT contact us earlier, as there is no immediate availability of replacement hardware. The original terms of purchase for warranty and support for the USB Host system will continue to be in place. Upon receiving your B0 motherboard, please return your old motherboard to Intel Corporation, Attn: Mike Givens, JF2-53, 2111 NE 25th, Hillsboro, OR 97124.

USB-enabled PC systems are now available commercially. If you need additional USB systems, please order them from the list of OEMs below or contact your local PC retailer for price and availability of these systems. The model numbers for the USB-ready systems are:

IBM - Model number 2176C6Y or higher

Compaq - Model number 4762 or higher

Toshiba - Model name Infinia

Siemens - please contact for model number

Sony - VAIO model PCV-70 or 90

Chapter 8:

You should make a point of visiting the Developers area of Intel's corporate web site at http://www.intel.com/design/usb for further technical updates on USB technology and Intel products. We look forward to seeing you at future compliance workshops and at Comdex this fall.

Sincerely,

Intel USB Team

Note: As of the printing of this book (February 1998) Beta 3 of Windows 98 has been released. The final version of Windows 98 is targeted for release in 2Q of this year. No more beta test participants are being accepted at this point.

9. Vendor List

9.1 Unsorted

(Except for this section, vendors are sorted by category.)

Acer

157, Shan-Ying Rd.

Kweishan, Taoyuan 333

Taiwan R.O.C.

886-3-359-5000

Fax: 886-3-359-5678

www.apl.com.tw

A-Com Computer BV

Kerkenbos 10-15 H.

6546 BB Nijmegen

The Netherlands

31-24 378 88 09

31-24 378 88 39

A-trend Technology Co., Ltd.

8862-698-2199

Fax: 8862-698-2369

Advanced-Connectek Inc.

3002 Dow Ave., #224

Tustin, CA 92680

(714) 573-1920

Fax: (714) 573-1924

ADI Corporation

14F, No. 1, Sec 4

Nanking E. Road

Taipei, Taiwan, R.O.C.

(886-2) 713-3337 x 5586

Fax: (886-2) 713-6555

Agiler Sysgration Ltd.

8 Fl., No. 542-7, Chung Cheng Rd., Hsin Tien,

Taipei, Taiwan, R.O.C.

(886-2) 218-2288

Fax: (886-2) 218-7663

www.sysgration.com.tw

Anchor Chips

12396 World Trade Drive

Suite 212

San Diego, CA 92128

(619) 676-6815

Fax: (619) 676-6896

sales@anchorchips.com

www.anchorchips.com

Unsorted

Behavior Tech Computer
4180 Business Center Dr.
Fremont, California 94538
(510) 657-3956
Fax: (510) 657-1859

Cable Systems International
505 North 51st Avenue
Phoenix, AZ 85043-2701
(602) 233-5645
Fax: (602) 233-5850

Chicony Electronics Co., Ltd.
No. 25, Wu-Gong 6th Rd.,
Wu-Ku Industrial Park,
Taipei Hsien
Taiwan R.O.C.
(886-2) 298-8120
Fax: (886-2) 298-8442

Digital Equipment Corporation
100 Nagog Park
AK01-1/C14
Acton, MA 01720-3499
(508) 264-6883
Fax: (508) 264-5854

Digital Stream
Santoku Bldg. No. 2-203
2719 Kamitsuruma
Sagamihara-shi
Kanagawa-ken 228 Japan
0427-47-0900
Fax: 0427-47-6011

ESS Technology, Inc.
48401 Fremont Blvd.
Fremont, CA 94538
(510) 492-1755
(510) 492-1237

EzKey Corp.
11F, No. 167, Fu Ho Rd.
Yung Ho City,
Taipei Hsien
Taiwan, R.O.C.
(886-2) 232-5838
Fax: (886-2) 232-5841
www.ezkey.com.tw

Fujitsu
5200 Patrick Henry Dr.,
Santa Clara, CA 95054
(408) 764-9327
Fax: (408) 982-9512

Future Technology Devices International Ltd.
St. George's Studios
93/97 St. George's Road
Glasgow G3 6JA

Chapter 9: Vendors

Scotland, UK

(44) 0141-353-3565

Fax: (44) 0141-353-2656

www.ftdi.co.uk

GVC Corporation

2F., No. 78, Sec. 2

An-Ho Rd.,Taipei

Taiwan R.O.C.

(886-2) 701-2226 x 332

Fax: (886-2) 704-0338

Hitachi America, Ltd.

1740 Technology Drive

Suite 420,

San Jose, CA 95110

(408) 436-3407

Fax: (408) 436-3414

www.hitachi.com

ITT Cannon

1851 East Deere Avenue

Santa Ana, CA 92705-5720

(714) 757-8459

Fax: (714) 757-8303

LG Electronics Inc.

184 kongdan-dong, Kumi-city,

Kyoungbuk, Korea

82-546-460-3266

Fax: 82-546-460-3272

LiteOn

5F, 16, Sec 4, Nanking E. Rd.,

Taipei, Taiwan, R.O.C.

(886-2) 570-6999 x345

Fax: (886-2) 570-6888

www.liteontc.com.tw

Longwell

21700 E. Copley Dr.

Suite 120

Diamond Bar, CA 91765

(909) 861-5527

Fax: (909) 396-6453

sales@longwell.com

www.longwell.com

Molex

641 Poplar Street

Orange, CA 92868

(714) 937-9380

Fax: (714) 978-9106

Northstar Systems

9400 Seventh Street

Bldg. A2

Rancho Cucamonga, CA

(909) 483-9900

Fax: (909) 944-0464

Unsorted

www.northstar.com

Oasis

7130 Engineer Road

San Diego, CA 92111

(619) 279-7400

Fax: (619) 279-0079

www.ocp.com

Pacetec IMG

The Boott Mills

100 Foot of John St.,

Lowell, MA 01852

(508) 970-0330

Fax: (508) 970-0199

www.spacetec.com

Samsung

416 Maetan-3Dong

Paldal-Gu,Suwon City

Kyungki-Do

Korea 442-742

(82-331) 200-7857

Fax: (82-331) 200-7233-4

Sejin Electron Inc.

60-19 Kasan-Dong Keumchon-Ku

Seoul 153-023, Korea

(02) 866-3333

(02) 864-3375

www.seijin.com

Shamrock Technology Co., Ltd.

7F No. 108-4 Min-Chuan Rd., Hsin-Tien City,

Taipei Hsien

Taiwan R.O.C.

(886-2) 218-2155

Fax: (888-2) 218-5154

www.shamrock-tech.com

SMC

6 Hughes

Irvine CA 92718

(714) 707-4835

Fax: (714) 707-2434

www.smc.com

Thrustmaster

7175 NW Evergreen Pkwy #400

Hillsboro, OR 97124

(503) 615-3200

Fax: (502) 615-3300

www.thrustmaster.com

Tyan Computer

1753 S. Main St.

Milpitas, CA 95035

(408) 956-8000 x120

Fax: (408) 956-8044

www.tyan.com

Universal Access

Aprilvagen 27 s-175 40

Jarfalla, Sweden

46 (0) 8-580-261-80

Fax: +46 (0) 8-580-308-21

USAR Systems

Bruce Tetelman

568 Broadway

New York, NY 10012

(212) 226-2042

Fax: (212) 226-3215

Sun Microsystems Computer Company

2550 Garcia Avenue

MS MPK14-201

Mountain View, CA 94043-1100

(415) 786-7032

Fax: (415) 786-7323

9.2 Integrated Circuits

Alcor Micro, Inc.

155A Moffet Place

Suite 240

Sunnyvale, CA 94089

(408) 541-9700

Fax: (408) 541-0378

Atmel Corporation

2325 Orchard Parkway

San Jose CA 95131

(408) 487-2605

Fax: (408) 487-2600

CAE Technology

2355 Old Oakland Rd., #41

San Jose, CA 95131

(408) 526-9270

Fax: (408) 526-9308

CMD Technology Inc.

1 Vanderbilt

Irvine, CA 92718

(800) 426-3832

(714) 454-0800

Fax: (714) 455-9554

Crescent Heart

3529 N.E. Bryce Street

Portland, OR 97212-1854

(503) 287-5422

Fax: (503) 287-5430

www.c-h-s.com/chschs

Cypress

3901 North First Street

San Jose, CA 95134

(800) 858-1810

Integrated Circuits

Fax: (408) 943-6848
www.cypress.com

FTDI
St. George's Studios
93/97 St. George's Road
Glasgow G3 6JA
Scotland, UK
(44) 0141 353 2565
Fax: (44) 0141 353 2565
www.ftdi.co.uk

Intel
CH6-214
5000 W. Chandler Blvd.
Chandler, AZ 85226
(602) 554-2897
Fax: (602) 554-1984
www.intel.com

1900 Prairie City Road,
Folsom, CA 95630
(916) 356-6079
Fax: (916) 356-2703
http://developer.intel.com/design/usb

Kawatsu
1735 Technology Drive
Suite 780
San Jose, CA 95110
(408) 467-1868
Fax: (408) 467-1878
usb@kawatsu.com
www.kawatsu.com

Lucent Technologies, Inc.
555 Union Boulevard
Room 30L-15P-BA
Allentown, PA 18103
(800) 372-2447
Fax: (610) 712-4106
www.lucent.com/micro

Mitsubishi
Electronic Device Group
Sunnyvale, CA
(408) 730-5900

Motorola
Motorola Semiconductors
Hong Kong Ltd.
13/F, Prosperity Centre, 77-81 Container Port Road
Kwai Chung, N.T.
Hong Kong
852-2612-5678
852-2612-5616
Fax: 852-2485-0548

MultiVideo Labs
29 Airpark Road
Princeton, NJ 08540

Chapter 9: Vendors

(609) 497-1930

Fax: (609 497-1945

www.mvl.com

National Semiconductor

2900 Semiconductor Drive

Mail Stop C2-633

Santa Clara, CA 95052-8090

(408) 721-8762

Fax: (408) 721-5691

NEC Electronics, Inc.

2880 Scott Blvd.

M/S SC2400

P.O. Box 58062

Santa Clara, CA 95052-8062

(408) 588-5436

Fax: (408) 588-6752

484, Tsukagoshi 3-chome, Saiwai-Ku

Kawasaki, Kanagawa 210, Japan

81-44-548-8858 Direct

Fax: +81-44-548-8889

NetChip Technology, Inc.

635-C Clyde Avenue

Mountain View, CA 94043

(415) 526-1490

Fax: (415) 526-1494

www.netchip.com

Philips Semiconductor

Interleuvenlaan 74-76

B-3001 LEUVEN-BELGIUM

32 16 390 643

Fax: +32 16 390 600

Philips Electronics North America Co.

Palo Alto, CA

(415) 846-4300

Philips Semiconductor

811 E. Arques Ave.,

Sunnyvale, CA 95066

408-991-3276

Fax: 408-991-2133

www.semiconductors.philips.com

ScanLogic Corporation

4 Preston Court

Bedford, MA 01730

(617) 276-3901

Fax: (617) 275-1758

Slinc@ix.netcom.com

Texas Instruments
Post Office Box 655303
Dallas, TX 75265
(972) 644-5580
Fax: (972) 480-7800
http://www.ti.com

Thesys
Haarbergstraße 61
D-99097 Erfurt, Germany
49 361 427 8350
Fax: +49 361 427 6161
www.thesys.de

Winond Electronics Corp.
No. 4, Creation Rd. III
Science-Based Industrial Park
Hsinchu, Taiwan, R.O.C.
886-3-5770066 ext. 7018
Fax: 886-3-5792646
www.winbond.com.tw

9.3 Power ICs

Micrel
1849 Fortune Dr.
San Jose, CA 95131
(408) 944-0800
Fax: (408) 944-0970
www.micrel.com

Texas Instruments
Post Office Box 655303
Dallas, TX 75265
972-644-5580
Fax: 972-480-7800
http://www.ti.com

Unitrode Corporation
7 Continental Boulevard
Merrimack, NH 03054
(603) 429-8504
Fax: (603) 429-8564
www.unitrode.com

9.4 Tools

Computer Access Technology Corporation
2403 Walsh Avenue,
Santa Clara, CA 95051-1302
(408) 727-6600
Fax: (408) 727-6622
sales@catc.com
www.catc.com

Digital Systems
Santa Ana, CA
(714) 838-2495

FuturePlus Systems

Chapter 9: Vendors

3550 N. Academy Blvd.
Suite 214
Colorado Springs, CO 80917-5088
(719) 380-7321
Fax: (719) 380-7362

Genoa
5401 Tech Circle
Moorpark, CA 93021
(805) 531-9030
Fax: (805) 531-9045
www.gentech.com

Intel
CH6-214
5000 W. Chandler Blvd.
Chandler, AZ 85226
(602) 554-2897
(602) 554-8080
Fax: (602) 554-1984
www.intel.com

1900 Prairie City Road
Folsom, CA 95630
(916) 356-6079
Fax: (916) 356-2703

Phoenix Technologies Ltd.
411 E. Plumeria Dr.

San Jose CA 95134
(408) 570-1000
Fax: (408) 570-1001

Keil Software
16990 Dallas Parkway
Suite 120
Dallas, TX 75248
800-348-8051
972-735-8052
Fax: 972-735-8055
sales.us@keilx.com
www.keil.com

PLC
Weatherford, TX
(817) 599-8363
www.plcorp.com

Tasking
Dedham, MA
(617) 320-9400

Thesys
Haarbergstraße 61
D-99097 Erfurt, Germany
49 361 427 8350
Fax: +49 361 427 6161
www.thesys.de

Xyratex
4101 Westerly Place

Cables

Suite 105
Newport Beach, CA 92660
(714) 476-1016
Fax: (714) 476-1916
www.xyrate.com

9.5 Help

Annabooks
11838 Bernardo Plaza Court
San Diego, CA 92128-2414
(619) 673-0870
Fax: (619) 673-1432
info@annabooks.com
www.annabooks.com

Doctor Design
10505 Sorrento Valley Road #1
San Diego, CA 92121-1608
(619) 824-3031
Fax: (619) 824-3131

Knowledge Tek
7230 W 119th Pl. Suite C
Broomfield CO 80020
(303) 465-1800
Fax: (303) 465-2600

MicTron

8610 17th Ave. N.E.
Seattle WA 98115
(206) 545-9449

Microsoft
www.microsoft.com/hwdev/wdmrsc.htm

Northstar Systems
9400 Seventh Street
Bldg. A2
Rancho Cucamonga, CA
(909) 483-9900
Fax: (909) 944-0464
USBSales@northstar1.com
www.northstar1.com

SystemSoft
2 Vision Drive,
Natick, MA 01760
(508) 651-0088
Fax: (508) 651-8188
usb@systemsoft.com/products/usb
www.systemsoft.com

USB News
www.usbnews.com

9.6 Cables

AMP

Harrisburg, PA

(800) 522-6752

CSI

1200 Johnson Ferry Rd., Suite 250,

Marietta, GA 30068-2798

(770) 579-0060

Fax: (770) 579-0063

Harting

2155 Stonington Ave.

Hoffman Estates, IL 60195-5211

(847) 519-7700

Fax: (847) 519-9771

Molex

641 Poplar Street

Orange, CA 92868

(714) 937-9380

Fax: (714) 978-9106

Newnex Technology Corp.

1190-T Miraloma Way

Sunnyvale, CA 94086

(408) 749-1480

Fax: (408)749-1963

sales@newnex.com

www.newnex.com

Northstar Systems

9400 Seventh Street

Bldg. A2

Rancho Cucamonga, CA

(909) 483-9900

Fax: (909) 944-0464

USBSales@northstar1.com

www.northstar1.com

9.7 Connectors

AMP

Harrisburg, PA

(800) 522-6752

Harting

Marcos A. Barrionuevo

2155 Stonington Ave.

Hoffman Estates, IL 60195-5211

(847) 519-7700

(847) 519-9771

Methode

Rolling Meadows, IL

(847) 392-3500

9.8 Keyboards

Alps Interactive

3553 N First Street

San Jose, CA 95134-1804

Systems

(408) 432-6000, 432-6503, 432-6570

Fax: (408) 321-8494

www.alps.com

Cherry Electrical Products

3600 Sunset Ave.,

Waukegan, Illinois 60087

(847) 360-3393, (847) 360-3500, (847) 360-3434, (800) 510-1689

Fax: (847) 360-3566

cep_sales@cherrycorp.com

www.industry.net/cherry.electrical

Key Tronic Corporation

P.O. Box 14687

Spokane, WA 99214-0687

(509) 928-8000, 927-5520

Fax: (509) 927-5503

info@keytronic.com

www.keytronic.com

9.9 Joysticks

Microsoft

One Microsoft Way

Bldg. Redwest West B/1062

Redmond, WA 98052-6399

(206) 703-2589

Fax: (206) 867-3548

9.10 Systems

AST

16215 Alton Parkway

Irvine, CA 92618-3618

(714) 727-4141

Fax: (714) 727-9355

www.ast.com

Compaq

Houston, TX

(713) 514-0484

www.compaq.com

Gateway 2000

610 Gateway Dr.

Sioux City, SD 57049

(605) 232-2000

Fax: (605) 232-2731

HP

(916) 785-5623

Colorado Springs, CO

(719) 590-1900

IBM

3039 Cornwallis Road

RTP, NC 27709

(919) 254-8465

Somer, NY
(800) 426-3333

(512) 838-6276
Fax: (512) 838-6292
www.pc.ibm.com

Sony
Park Ridge, NJ
(201) 930-1000

Toshiba America Information Systems, Inc.
9740 Irvine Blvd.
Irvine, CA 92618-1697
(714) 583-3000
Fax: (714) 583-3893
www.toshiba.com

Award Software International
777 East Middlefield Road,
Mountain View, CA 94043-4023
(415) 968-8886
(415) 968-4433
Fax: (415) 968-0274
www.award.com

American Megatrends, Inc.

6145-F Northbelt Parkway
Norcross, GA 30071-2976
(770) 264-8600
Fax: (770) 246-8791
www.megatrends.com

9.11 Mice

Alps
3553 N First Street,
San Jose, CA 95134-1804
(408) 432-6000(main)
432-6503, 432-6570
Fax: (408) 321-8494
Pager: 800-509-6717
www.alps.com

KYE
2605 E. Cedar Street
Ontario, CA 91761
(909) 923-3510
(800) 456-7593
Fax: (909) 923-1469

Logitech
Fremont, CA
(510) 795-8500

9.12 Printers

Cannon

3-30-2 Shimomaruko Ohta-ku

Tokyo 146 Japan

81-3-3758-2111

Fax: 81-3-3756-6052

www.canon.com

9.13 Computer Telephones

Mitel

350 Legget Drive,

Kanata, ON, Canada, K2K 1X3

(613) 592-2122

(800) MITEL-SX

Fax: (613) 592-4784

www.mitel.com

Nortel

(408) 565-7454

Fax: (408) 565-8080

9.14 Video Cameras

Compaq

PO Box 692000

Houston, TX 77269-2000

(713) 514-9542

Fax: (713) 514-0924

www.compaq.com

Vista Imaging, Inc.

521 Taylor Way

Belmont, CA 94002

(650) 802-9685

Fax: (650) 802-0322

www.vistaimaging.com

Xirlink

2210 O'Toole Ave.

San Jose CA 95131

(408) 324-2100

(408) 324-2101

www.xirlink.com

9.15 Infra-Red

e^{TEK} Labs

1057 E. Henrietta Rd.,

Rochester, NY 14623

(716) 292-6400

Fax: (716) 292-6273

etek@vivanet.com

www.eteklabs.com

9.16 Motherboards

American Megatrends, Inc.
6145-F Northbelt Parkway
Norcross, GA 30071-2976
(770) 264-8600
Fax: (770) 246-8791
www.megatrends.com

Diamond Flower
135 Main Avenue,
Sacramento, CA 95838
(916) 568-1234
Fax: (916) 568-1233
Info@dfiusa.com
http://www.dfiusa.com

Intel
CH6-214
5000 W. Chandler Blvd.
Chandler, AZ 85226
(602) 554-2897
(602) 554-8080
Fax: (602) 554-1984
www.intel.com

1900 Prairie City Road
Folsom, CA 95630
(916) 356-6079

Fax: (916) 356-2703

9.17 Software Core

Cicada Semiconductor Inc.
1717 W. 6th St., Suite 440
Austin, TX 78703
(512) 703-1633
Fax: (512) 703-1716
nvb@cicada-semi.com

Oki Semiconductor
785 North Mary Ave.
Sunnyvale CA 94086-2909
(408) 720-1900
Fax: (408) 720-1918

Phoenix
(408) 535-2528

Santa Clara, CA
(408) 654-9000
www.ptltd.com

Sand Microelectronics, Inc.
3350 Scott Blvd., #24
Santa Clara, CA 95054
(408) 235-8600
Fax: (408) 235-8601
www.sandmicro.com

Decicon Inc.

1250 Oakmead Parkway
Suite 316
Sunnyvale, CA 94086
(408) 720-7690
(408) 720-7691

9.18 Gamepad

Microsoft
Redmond, WA
(206) 882-8080

9.19 USB Host and Function Macros in Netlist or RTL USB Test Environment

Sapien Design
45335 Potawatami Dr.
Fremont, CA 94539
(510) 668-0200
Fax: (510) 668-0200
sapien@pacbell.net
www.sapeindesign.com

9.20 Standards

Underwriters Lab, Inc.
1655 Scott Blvd.

Santa Clara, CA 95050
(408) 985-2400

Vesa
2150 North First Street
Suite 440
San Jose, CA 95131-2029
(408) 435-0333
Fax: (408) 8225

Institute of Electrical and Electronic Engineers
445 Hoes Lane
P.O. Box 1331
Piscataway, NJ 08855-1331
800-678-4333
(908)-981-9667

Federal Communications Communication (FCC) Technical Standards Branch
2000 M St. NW
Washington, DC 20554
(202) 739-0704
Fax: (202) 887-0198

10. Index

1284 Parallel Port, 61
1394, 18
1TR6, 54
5ESS, 54
68PM302, 43
8051, 93
8x930, 122
8x930 Ax, 185
8x930 Hx, 193
A to B Cable, 30
Access.bus, 283
Acer, 311
ACK, 283
acknowledgment, 283
A-Com Computer BV, 311
ACPI, 18
Active Device, 283
ADB, 283
address, 176
ADI Corporation, 311
Advanced Configuration and Power Interface, 18
Advanced Power Management, 283
Advanced-Connectek Inc., 311
Agiler Sysgration Ltd., 311
Alcor Micro, Inc., 91, 130, 315
Alps, 45, 323
Alps Interactive, 321
American Megatrends, Inc., 19, 127, 323, 325
AMP, 320, 321
analog transceiver, 163
analyzer, 175
Anchor Chips, 92, 311
Annabooks, 7, 24, 232, 320
ApBUILDER, 127
APM, 283
Apple Desktop Bus, 283
application logic, 199
application simulation model, 198
arbitration, 141

Ashley Laurent, Inc., 19, 178
ASIC, 37, 87
assembly software, 180
AST, 322
asynchronous, 119
asynchronous data, 283
asynchronous messaging, 6
Atmel Corporation, 132, 315
A-trend Technology Co., Ltd., 311
audio, 6, 18, 21, 115, 164
Audio Device Class, 25
Austel, 54
auto-launch, 187
AVM, 53
Award, 19
Award Software International, 19, 323
babble, 283
back-to-back transfer, 107
bandwidth, 187
battery capacity, 162
battery performance, 18
Behavior Tech Computer, 312
bi-directional communication, 167
bi-directional parallel interface, 118
big endian, 284
BIOS, 54, 56
bit stuff error, 176, 203
bit stuffing, 108, 126, 149, 215, 221, 284
blocks, 35
books, 232
boundary condition, 199
buffer, 79
build number, 17
bulk capacitance, 274
bulk data transfer, 119
bulk in pipe, 169
bulk out pipe, 169
bulk transfer, 125, 150, 284
bus analyzer, 175

327

bus arbitration, 141
bus enumeration, 284
bus fault, 74
bus master, 141
bus reconfiguration, 74
bus topology, 188
bus voltage, 86
bus voltage, isolation, 86
bus-powered, 76
bus-powered device, 84, 250
bus-powered function, 235
bus-powered hub, 83, 235
bus-powered hub requirements, 269
buttons, 23
bypass capacitors, 273
byte boundary error, 204
cable, 40
cable characteristics, 34
cable impedance, 33
cable length, 33
cable resistance, 275
Cable Systems International, 312
cables, 28, 77, 85, 123, 251, 274, 320
CAE Technology, 315
camera, 20, 45
CAN, 43
Cannon, 324
capacitance, 161
capacitance, bulk, 163
capacitor, 264
capacitor charge current, 271, 272
CAPI, 54
CE, 75, 77
channel administration, 142
channel buffer, 143
chassis ground, 252
Cherry Electrical Products, 322
CHI, 284
Chicony Electronics Co., Ltd., 312
chip set, 36
CIF, 116
Class Driver, 24
class/minidriver structure, 21
client, 284
client software, 24, 36, 37
clock, 79

clock error, 204
clock multiplication, 99
CMD Technology, 60, 92, 96, 97, 181, 315
codec, 115
color scanner, 64
command packet, 169
CommMod, 43
Common Class, 22, 25
Common Class Specification, 18
CommSock, 43
communications, 19, 21
Communications Device Class, 25
communications manager, 43
Compaq, 2, 26, 220, 298, 322, 324
comparator, 117
compatibility, 85
compatibility workshop, 76, 79
compliance, 76, 183
component object model, 17
compound device, 75
compressed video, 6
compression, 116
Computer Access Technology Corporation, 127, 318
computer telephones, 324
conferencing, 17
configuration, 24, 77, 79, 94
configuring software, 285
connect event, 271
connectivity, 74
connector, 27, 40, 85, 264, 275, 305, 321
connector resistance, 275
consultants, 24
control pipe, 56, 168, 285
control transfer, 150, 285
controller, 76, 83
core logic, 87
CotoWabash, 245
CRC, 91, 108, 126, 131, 146, 149, 152, 176, 203, 215, 285
CRC coding, 221
Crescent Heart, 315
CSA, 40, 75, 77, 86
CSI, 321
CT1, 54

CTI, 2, 285
current capacity, 81
current draw, 81
current limit, 238, 241, 263
current limit device, 243
current limiting, 162, 245, 254, 279
current sensing, 91, 131
cyclic redundancy check, 285
Cypress Semiconductor, 99, 121, 134, 180, 315
DAC, 164
data acquisition, 196
data conversion, 108
data encoding, 108, 215
data pattern, 176
data recovery, 228
data size, 24
data transfer, 119, 185
data transmission, 163, 164
DCOM, 17
DD, 25
DDK, 20, 73
debug, 117
debug software, 180
debugging, 196
DEC, 2, 283, 298
Decicon Inc., 204, 231, 325
decoding, 108
decoupling, 133
default pipe, 285
design goals, 72
design information, 232
design service, 197
Detroit, 19, 36
developer's conference, 232
developer's kit, 181
development tools, 127
device, 27, 36, 249, 285
Device Bay, 15
device class, 18, 235, 249
Device Class, 25
device class specifications, 21
device configuration, 79, 188
device controller, 91, 106, 134
device design, 77
device driver, 25, 72, 178, 182
device driver development, 195

device failure, 238
device firmware, 91
device IC, 36, 79, 116, 122
device information, 24
device model, 203
device synthesizable core, 223
device, bus-powered, 84
devices, self-powered, 83
Diamond Flower, 325
differential signal, 33, 85
digital audio, 18
Digital Equipment Corporation, 312
Digital Stream, 312
Digital Systems, 318
Digital Versatile Disk, 18
digital video camera, 20, 45
distributed component object model, 17
DLL, 185
DMA, 118, 150
DMA bridge, 141
DMI, 286
docking station, 253
Doctor Design, 320
DOS, 195
downstream, 26
downstream ports, 76
driver, 20, 21
driver development, 17
drivers, 36, 37
droop, 237, 258
drop-out voltage, 239, 240
duplex audio, 115
DVD, 18
ECP, 167, 192
ECU, 67
E-DSS1, 54
EIDE controller, 121
electrolytic capacitor, 161
Elite, 19
embedded port, 76
EMC compliance, 304
EMI, 86, 132, 248, 267, 277
EMI/RFI, 40
encoding, 108, 215
endpoint, 24, 125, 141, 150, 157, 176, 205, 216, 218, 224, 229, 287

329

endpoint descriptor, 227
endpoint management, 140
Enhanced Video Adapter, 35
enumeration code examples, 23
EOF, 287
EOP, 108, 287
EOP error, 204
EPP, 167, 192
error, 176
error and transaction logging, 204
error condition, 203
error detection, 76, 79
error flag, 263
error flag output, 280
error handling, 107, 119
error insertion and detection, 198
error recovery, 225
ESS Technology, Inc., 312
e^{TEK} Labs, 59, 62, 195, 324
Ethernet, 43
EVA, 35
event counter, 117
EzKey Corp., 312
fan-out, 26
FAQ, 296
FAT32, 17
fault detection, 119
fault recovery, 107
fax, 64
FCC, 86
FCC class, 75, 76, 77, 79
Federal Communications
  Communication (FCC) Technical
  Standards Branch, 326
ferrite bead, 133, 258, 264
ferrite core, 86
FIFO, 79, 108, 125, 215
file system, 17
filter, 264
filtering, 79
fire hazard, 86
FireWire, 15, 18, 287, 300
firmware, 95, 131, 135, 137, 180, 183
flash BIOS, 54
flow control, 107
flyback voltage transients, 273
frame, 24, 287

Fs, 287
FTDI, 316
Fujitsu, 312
full-duplex, 287
Fully Rated Cable, 30
function, 287
function controller, 83
Function IC, 120
functional blocks, 35
fuse, 233, 243, 244, 254
Future Technology Devices
  International, 102, 134, 312
FuturePlus Systems, 175, 318
game controller, 19
gamepad, 46, 326
gang mode power, 270
ganged ports, 237, 242
Gateway 2000, 58, 322
Genoa Technology, 195, 319
GeoPort, 287
ground loop current, 252
ground loops, 133
ground plane, 161, 278
grounding, 85
GVC Corporation, 313
H.32x, 116
handshake, 225, 283
handshake error, 204
handshake packet, 287
handshaking, 25, 76, 79
hardware components, 25
hardware emulator, 180
hardware quality lab, 23
Harting, 321
HCD, 24
HCI drivers, 305
heat dissipation, 259
HID, 19, 22, 25, 97, 181
HID Class Support, 23
high side switch, 162
high-power function, 250
Hitachi America, Ltd., 313
host, 2, 25, 249, 287
host controller, 25
Host Controller, 36
host controller core, 226
host controller development kit, 181

Host Controller Driver, 24
host IC, 121
host model, 202
host simulation model, 198
host-side software, 183
hot plug, 56, 237, 267, 273
Hot Plug and Play, 119, 164
hot swapping, 175
hot-attach, 258
hot-plug, 240
hot-swapping, 63
HP, 322
HP logic analyzer, 175
hub, 19, 21, 24, 26, 66, 123, 129, 183, 226, 249, 288, 306
hub and monitor controller, 151
hub chip, 132
Hub Class, 25
hub configuration, 74, 76
hub controller, 134, 139, 146, 159
hub design, 74
hub device, 134
hub functionality, 74
hub IC, 74, 75, 148, 153, 154, 207
hub support, 193
hub synthesizable core, 228
hub, bus-powered, 83
hub, monitor, 5
hub, self-powered, 82
hub, stand alone, 4
Human Interface Device, 19, 22
Human Interface Device Class, 25
I/O request packet, 24
I^2C, 93, 116, 151, 159, 184, 283, 288
I^2C defined, 75
I_2O, 15
IBM, 2, 298, 322
IC, 36
IC compliance, 76
ICM, 27
idle, 124
IEEE 1394, 15, 287, 288, 300
IEEE-1284, 166, 170
IEEE-1394, 18, 21
image, 19
image capture, 115
Image Class, 22

Image Class Driver, 19
image compression, 116
Image Device Class, 25
impedance, 33
In-Circuit Emulators, 127
industrial design, 87
Industry Standard Architecture, 288
infra-red, 324
infrared remote control, 62
Ing Buro H. Doran, 52, 67
Ing. Buro H. Doran, 43
initialization, 81
inrush control, 272
inrush current, 163, 238, 258, 271
inrush current limiting, 245
Institute of Electrical and Electronic Engineers, 326
In-System Design, 61
integrated circuits, 315
Integrated Services Data Network (ISDN), 288
Intel 930, 120
Intel Corporation, 2, 22, 26, 122, 294, 298, 304, 307, 316, 319, 325
Intel MMX, 18
intellectual property, 301
Inter Integrated Circuit, 75
interconnect hub, 251
interface, 157
interface controller, 150
Interlayer Communication Model, 27
Internet conferencing, 17
Internet Explorer 4.0, 17
interoperation, 76, 79
inter-pipe synchronization, 170
interrupt, 117, 125, 142, 150, 157, 226, 288
interrupt pipe, 169
interval timer, 117
IR remote control, 52
IrDA, 135
IRP, 24
IRQ, 142, 289
ISA, 177, 288
ISDN, 17, 53
ISO 9000, 40

331

ISO9141, 67
isochronous, 119, 125, 289
isochronous data transfer, 6
isochronous pipes, 107
isochronous sources, 77
isochronous transfer, 150
iSYSTEMS, 127
ITT Cannon, 313
jitter, 289
joystick, 102, 322
joystick controller, 104
junction temperature, 161
Kawatsu, 136, 316
Keil Software, 319
Key Tronic Corporation, 55, 322
keyboard, 19, 25, 55, 120, 135, 268
keyboard controller, 97, 102, 121
keyboard developer's kit, 180
keyboards, 321
Knowledge Tek, 320
KwiKey IR, 62
KYE, 323
LAN, 15
laptop, 253
latency, 107, 141
legacy support, 305
LG Electronics Inc., 313
linear regulator, 256
Linear Tech, 245
LiteOn, 313
little endian, 289
LOA, 289
load, 248, 253
loading, 238, 240
logic analyzer, 178, 196
logical device, 27
Logitech, 63, 323
Longwell, 313
low-power function, 251
low-speed interface, 157
Lucent Technologies, Inc., 106, 141, 166, 316
market, 8
mass storage, 19, 22
Mass Storage Device Class, 25
master, 2
Maxim, 245

maximum length, 33
MCA, 290
mechanical design, 87
Memphis, 17, 36, 65, 72
message pipe, 290
Metalink Corporation, 127
Methode, 321
mice, 135, 323
Micrel, 162, 245, 247, 256, 257, 279, 318
microcontroller, 76, 79, 113, 117, 121, 123, 147, 158
microcontroller core, 99
microprocessor, 118
Microsoft, 2, 26, 72, 220, 290, 298, 322, 326
Microsoft Developer Network, 73
MicTron, Inc., 197, 320
minidriver, 21
missing frame, 176
Mitel, 44, 324
Mitsubishi, 111, 316
MMX, 18, 45, 115
modem, 17, 250, 290
Molex, 313, 321
monitor, 19, 22, 252
monitor communication IC, 145
monitor control, 183
monitor controller, 146
Monitor Device Class, 25
monitor filament voltage, 161
monitor hub, 5, 76
monitor hub controller, 136
monitor model, 204
MOSFET switch, 233, 245
motherboard, 36, 58, 123, 325
Motorola, 43, 113, 316
mouse, 19, 25, 101, 120
mouse controller, 121
mouse microcontroller, 113
mouse/joystick controller, 104
MS Word 6.0/95 Binary Converter, 23
MS Word Viewer 97, 23
MSDN, 20, 22, 73
multimedia, 164
multiple isochronous sources, 77

multiplexer, 144
MultiVideo Labs, 145, 316
NACK, 290
National Semiconductor, 26, 146, 220, 317
NDIS 4.0, 59
NEC, 2, 298, 317
NetChip Technology, Inc., 148, 150, 317
Newnex Technology Corp., 321
NI1, 54
Nogatech, Inc., 115
Nohau Corporation, 127
noise, 85
noise immunity, 34, 76, 79
non-isochronous pipes, 107
Nortel, 2, 324
Northern Telecom, 298
Northstar Ssytems, 320
Northstar Systems, 27, 313, 321
Northstar Systems, Inc., 40, 66
notebook computer, 48
NRZI, 91, 108, 126, 131, 149, 215, 290
NRZI encoding, 221
Oasis, 314
OEM Service Release, 19
OHCI, 19, 85, 290, 303, 306
OHCI host synthesizable core, 220
OHCI registers, 227
OHCI, defined, 26
Oki Semiconductor, 206, 212, 325
OLE, 17
OnNow, 18, 21
on-resistance, 162, 241
Open HCI, 220
Open HCI Specification, 220
Open Host Controller Interface, 19, 290
Open Host Controller Interface (OHCI), 26
operating system, 17, 21, 56, 72
Opti Semiconductor, 26
OS Group, 19
OS/2, 195
OSR, 17, 19
OSR 2.1, 299

OSR2, 186
output filter, 264
overcurrent, 245
overcurrent detection, 238
overcurrent limiting device, 254
overcurrent protection, 81, 230, 233
overcurrent trip, 254
p.c. board resistance, 242
Pacetec IMG, 314
package protocol sequencing, 108
packet, 168, 290
packet ID, 126
packet sequencing, 126
packets, 37
parallel conversion IC, 166
parallel interface, 118
parallel port, 167
parallel port device conversion, 191
parallel port register, 171
parallel printer, 61
parallel to serial conversion, 221
PBX, 53, 290
PC chassis, 15
P-channel MOSFET switch, 272
PCI, 2, 177, 220, 226
PCI to USB add-in card, 182
PCI-ISA bridge, 121
PCI-to-USB controller board, 60
PCI-to-USB controller chip, 96
PCMCIA, 291
PC-to-PC connectivity, 59
PDK, 307
performance, 76
peripheral, 27, 93
peripheral controller, 121
Peripheral Developer Kit, 307
peripheral use per host, 8
peripherals, 36, 298
phase, 291
Philips Semiconductor, 64, 151, 163, 283, 317
Phoenix Technologies, 127, 187, 198, 216, 220, 319, 325
PID, 126, 152, 176, 215, 291
PID error, 204
pig tail, 28
Pig-Tail Cable, 33

pipe, 24, 168, 291
PLC, 319
PLL, 291
plug, 28
Plug and Play, 3, 18, 21, 137, 164, 183, 187, 191, 193, 291, 298
Plugfest, 7, 22, 37, 72, 79, 85, 223, 228
PMC, 177
PnP, 291
polling, 291
polyfuse, 233, 243, 254
Polyfust, 243
PolySwitch, 243
POR, 291
port, 291
port connect/disconnect, 74
port limit, 253
port state, 74
port status, 133
ports, 76
POTS, 288, 291
power, 22, 77, 239
power and ground traces, 258
power control, 160
power design example, 264
Power Device Class, 25
power dissipation, 259
power distribution, 233, 236
power down, 124
power ICs, 318
power interface, 245
power loads, 269
power management, 18, 21, 80, 100, 248
power on reset, 291
power regulator, 256
power supply, 253
power supply voltages, 239
power switch, 162, 257, 272
power switching, 75, 131, 229, 230
power utilization, 188
power wire, 253
power-saving mode, 124
preprocessor, 176
price, 299

printed circuit board layout, 258, 277
printer, 19, 22, 253, 324
printer cable, 61
printer conversion, 191
Printer Device Class, 25
product development service, 195
products, 7
protocol analyzer, 195
protocol control, 141
protocol violation, 177
PS/2, 120, 135
PS/2 keyboard, 222
PS/2 mouse, 97, 102, 104
PTC resistor, 233
pull-up resistor, 83, 273
QCIF, 116
RA, 292
rate adaptation, 292
Rate Adaptation, 283
real-time compare, 124
real-time data, 6
real-time-clock, 121
receive FIFO, 110
receptacle, 28
Reciprocal Covenant Agreement, 301
re-initialization, 233
remote wake-up, 94
repeater, 74, 76
repeater state, 295
request, 292
request packet, 24
reset, 270
resistance, p.c. traces, 278
resistor, 78, 83
resume mode power, 271
retire, 292
RFI, 40
RISC, 99
root, 292
root hub, 24, 226, 306
RS-232, 2, 52, 68, 184
safety, 86
sample rate (Fs), 292
Samsung Semiconductor, Inc., 116, 314

Sand Microelectronics, Inc., 201, 223, 226, 228, 325
Sapien Design, 326
ScanLogic Corporation, 118, 317
scanner, 19, 20, 64
scanner control interface, 165
SCSI, 15, 292
Sejin Electron Inc., 314
self-powered, 76
self-powered device, 83, 249
self-powered function, 250
self-powered hub, 82, 235, 238, 250, 252
self-powered hub requirements, 253
serial data transmission, 163, 164
serial interface engine (see also SIE), 25
serial to parallel data conversion, 221
serial to parallel decode, 175
Series A and B connectors, 28
Series A Cable Plug, 28
Series A Receptacle, 31
Series B Cable Plug, 29
Series B Receptacle, 32
service, 292
Shamrock Technology Co., Ltd., 314
shielding, 33, 85
short, 238
short-circuit limit, 256
SIE, 25, 36, 91, 99, 108, 110, 126, 131, 140, 149, 164, 215, 221, 225, 229, 303
SIE/transceiver, 76, 79
signal termination, 273
signaling mode, 123
signaling violation, 177
SIMM, 177
simulation model, 201
slave, 2
SMC, 314
SOF, 293
SOF token, 227
SOF token generation, 221
soft start, 238, 272
soft start circuit, 163
software core, 325

solder connection, 258
solid state switch, 233, 243, 245, 254
Sony, 323
SOP, 108
speakers, 64
speed, 33, 76, 77, 79
speed, USB, 3
SPI, 293
SRC, 293
stage, 293
stand-alone hub, 66
standards, 326
state machine, 76, 231
state management, 24, 74
stereo D/A bitstream converter, 164
still image, 19
still image camera, 20
Stream Class, 20
stream pipe, 293
Sub-Channel Cable, 30
Sub-QCIF, 116
Sun Microsystems Computer Company, 315
suspend, 283
suspend mode, 81
suspend mode power, 270
suspend power, 273
suspend/resume, 74, 76
switched power, 269
Symbios, 26
synchronization type, 293
synchronous RA, 293
synchronous SRC, 293
synthesizable core, 216, 220, 223
synthesizable function core, 204
system drivers, 180
system loading, 240
systems, 322
SystemSoft Corporation, 19, 73, 127, 183, 191, 320
tantalum capacitor, 264
target state, 177
Tasking, 319
TDM, 293
Teledyne, 245
telephone, 2, 44
telephony, 115, 123

terminating, 123
termination, 273, 293
testing, 85
Texas Instruments, 153, 154, 245, 318
the interlayer communication model, 35
thermal dissipation, 277
thermal limit, 238
thermal limiting, 245
thermal management, 259
thermal overload, 260
thermal protection, 161, 234, 279
thermal runaway, 255
thermal shutdown, 162, 163, 263
Thesys, 157, 318, 319
throughput, 76
Thrustmaster, 314
timer, 117
timing analysis, 177
TIO, 117
token, 108, 294
token error, 204
tools, 318
topology, 4
Toshiba, 22, 48
Toshiba America Information Systems, Inc., 323
touch screen, 120
touch screen developer's kit, 180
traffic generator, 77
training seminars, 24
transaction, 294
transceiver, 25, 76, 79, 91, 99, 118, 123, 127, 131, 132, 140, 146, 152, 231
transfer, 294
transfer rate, 6
transfer type, 24
transfer types, 107, 123, 126
transient droop, 258
transmit FIFO, 109
transport channel, 143
turnaround time, 294
TUV, 86
TWAIN interface, 119
twisted pair, 33

Tyan Computer, 314
UHCI, 19, 85, 294, 303, 306
UHCI specification, 23
UHCI, defined, 26
UL, 40, 75, 77, 86, 236
UltraSCSI, 15
undervoltage lockout, 163
Underwriters Lab, Inc., 326
unit load, 248
Unitrode Corporation, 159, 318
Universal Access, 315
Universal Host Controller Interface, 19, 294
Universal Host Controller Interface (UHCI), 26
UNIX, 195
upstream, 26, 295
upstream cable, 273
USAR Systems, 120, 180, 315
USB, 298
    defined, 2
USB and PCI relationship, 3
USB blocks, 35
USB Class Driver, 19, 24
USB components, 13
USB Core Specification, 21
USB device class, 18
USB Device Driver, 25
USB features, 14
USB goals, 6
USB Host Controller Driver, 24
USB hub, 3
USB Image Class, 22
USB Implementers Forum, 301
USB market, 8
USB PC back panel, 15
USB PDK, 307
USB port, 3
USB products, 7
USB speed, 33
USB System Operation Summary, 37
USB topology, 4, 123
USB/PnP white papers, 23
USBD, 24, 295
USB-IF, 6, 76, 85, 301
USB-IF home page, 302

usbnews, 24
USB-to-ECU bridge, 67
USB-to-parallel conversion IC, 166
USB-to-RS232 bridge, 68
utility software, 187
vendor ID, 304
verification, 201
Verilog, 198, 201, 217, 223, 226, 228
Vesa, 326
VESA, 35
VHDL, 164, 201, 303
video, 6
video camera, 20, 45, 324
video chip, 115
video conferencing, 45, 115
video display, 252
Video Phone, 45
Virtual Chips, 198, 217, 223
virtual device, 295
Vista Imaging, Inc., 324
Visual C++, 22
VLSI, 61
VLSI design, 197
VME, 177
VN3, 54
voice, 6
voltage budget, 238
voltage distribution, 237
voltage droop, 240
voltage drop, 81, 237, 257, 260, 261, 263, 274, 278
voltage measurement, 261
voltage regulation, 233, 248
voltage requirements, 240

voltage transients, 273
VXI, 177
wake-up, 94
wall transformer, 82
WDM, 18, 19, 20, 21, 23, 73, 183, 186, 191, 193
web sites, 22
Win32 Device Module, 73
Win32 Driver Model, 18, 21, 183, 186
Winbond Electronics Corp., 121, 159, 165
Windows, 195, 299
Windows 95, 19, 54, 183, 186, 193, 195, 220
Windows 95 Gold, 19
Windows 95 utility software, 187
Windows 98, 17, 22, 26, 309
Windows hardware quality lab, 23
Windows NT, 20, 22, 183
WinHEC, 6, 73
Winond Electronics Corp., 318
wire gauge, 274
Word 6.0/95 Binary Converter, 23
Word Viewer 97, 23
workshop, 232
write pointer, 110
www.usb.org, 6
Xirlink, 45, 324
Xyratex, 319
ZAW, 18, 21
Zero Administration Initiative for Windows, 18
Zero Administration Windows, 21

Notes

Notes

Notes

Notes

Other USB Books from Annabooks

USB Handbook
By Kosar Jaff

Developing USB PC Peripherals
By Wooi Ming Tan

USB Hardware and Software
(Coming Soon)

By John Garney

Ed Solari

Shelagh Callahan

Kosar Jaff

Brad Hosler

Also available from Annabooks University

USB Design Workshops and Conferences

Annabooks

800-462-1042

619-673-0870

619-673-1432 FAX

info@annabooks.com

http://www.annabooks.com